DR. SMARTPHONE: AN ETHNOGRAPHY OF MOBILE PHONE REPAIR SHOPS

NICOLAS NOVA & ANAÏS BLOCH
2020

"DR. SMARTPHONE"
"MOBILE CITY CENTER"
"DOCTEUR IT"
"IKLINIK"
"LA CLINIQUE DU TÉLÉPHONE CELLULAIRE"
"BOXREPAIR"
"KING OF GSM"
"PHONETIME"
"IPHONE CLINIQUE"
"SMARTPHONE CLINIQUE"
"GOLDPHONE"
"PHONETIME"
"PHONE SERVICES"
"CYBER PHONE"

(not to be mistaken with "cyberphone," two streets away)... These are some of the names of a new type of business that has appeared in towns and villages in the past ten years. Their storefronts are easily recognisable, with signs invoking medical metaphors and the tropes of digital culture (with "cyber" or "i-" prefixes). Their services include

"BROKEN SCREEN REPAIR"
"PHONE FLASHING"
"BATTERY REPLACEMENT"
"WET PHONE RESCUE"
"SIM CARD UNLOCKING"
"BUTTON/CAMERA REPLACEMENT"
"DATA RECOVERY"
"WE UNLOCK."

Smartphones are dropped, their screens crack, buttons are damaged, dust builds up in the handset's many ports and openings. Sometimes the problem transcends the phone's hardware: an app needs updating, the operating system is wrongly configured, or the user wants to make space in the device's overloaded memory.

These spaces, and the people who run them, are part of an emerging market in the maintenance and repair of smartphones and other technological artifacts. Commercial stores represent the most visible element of this phone repair ecosystem, but similar practices can be seen in hackerspaces, Fab labs, and temporary venues such as repair cafés[1]—spaces where people come together to socialize, collaborate, and share knowledge.[2] Though there is some overlap, the main difference between hackerspaces and Fab labs is cultural. Hackerspaces are community-operated open laboratories that incorporate elements of machine shops, enabling people to come together to share resources and make things. Fab labs are a network of spaces that provide access to a core set of tools—including basic electronics equipment, laser cutters, and 3D printers.[3] Such spaces can and do support practices of repair and maintenance, but this is not their primary purpose, being geared more towards tinkering, making, and open hardware projects.

This book focuses on these stores and spaces, and the practices of the smartphone repairers who occupy them. How do these individuals come to end up fixing users' devices? How do they learn to handle products that were not designed to be repaired? And what can the mending of a cracked phone display tell us about skill, innovation, and the use of technology?

AGAINST DIGITAL OBSOLESCENCE

The rapid spread of these repair spaces poses questions about the sustainability of contemporary technical objects. The smartphones repaired, refurbished, and jailbroken in these spaces are, elsewhere, held up as icons of a continuous innovation regime (Neff & Stark, 2003) based on a rapid turnover of technological products and services. Some claim that this regime depends on manufacturers designing for the "planned obsolescence"[4] of their

1 A repair café is an informal meeting, supplying tools and materials to help people repair their possessions.

2 See Gershenfeld (2011); Anderson (2012); Cavalcanti (2013). See Davies (2017) for a more critical perspective on such spaces, and Richterich & Wenz (2017) for an analysis of the terminology.

3 Launched by researchers from the Center for Bits and Atoms at the MIT Media Lab, their charter requires them to be open to the public for little or no cost, and as such, most are run by local non-profit organisations.

4 Coined by Bernard London in 1932, "planned obsolescence" refers to techniques designed to reduce a product's lifespan in order to induce its replacement. Commonly cited examples include the heating elements of hair dryers or toasters, which allow these objects to be cheaply and easily produced, albeit with a shorter lifespan. Similarly, Apple's iPhone and MacBook batteries are accused of being difficult to modify or upgrade, and their repair is often discouraged. See Slade (2006)

goods, though the extent to which this has been an intentional business strategy remains a subject of ongoing debate.[5] As a result, manufactured products, often designed to last only a few years, are rarely repaired. Simultaneously, the high turnover of many of today's technological products is a key factor in the current environmental crisis; contributing to rising levels of carbon emissions, soil and air pollution, biodiversity loss, energy consumption, and the exploitation of non-renewable resources.
Recently, the short lifespan of smartphones has attracted criticism and pushback, with governments legislating to support repair practices,[6] and mobile phone manufacturers—both large incumbent companies and newer, smaller firms, such as Fairphone[7]—taking steps to make it easier to repair their hardware. Simultaneously, a secondary repair ecosystem has appeared independent of any action by smartphone manufacturers, with repair and maintenance facilities present, for example, in hackerspaces, for-profit stores, and temporary repair cafés. In this book, we argue that a greater attention to these spaces and repair practices can reveal how users are trying to handle the unsustainability of their digital devices.

Confronted with a broken smartphone, owners have several possible options: (1) if the device is covered by an active warranty, engage with the phone contract provider or manufacturer's customer service; (2) seek support from an approved third-party repairer, certified by the manufacturer; (3) go to an unauthorised local repair store; or (4) handle the issues themselves, using online documentation and tutorials, or workshops offered by non-profit institutions, like the hackerspaces and repair cafés described above. In these spaces, repair practices may include replacing broken screens, buttons or batteries, drying the device

for a more recent perspective on the activities of American companies, or Gabrys (2013), who approaches the debate around obsolescence by looking at circuits and spaces of electronics disposal.

5 See Vance Packard (1960), who further develops the idea by introducing a typology comprising obsolescence of function, obsolescence of quality, obsolescence of desirability, and psychological obsolescence.

6 See, for example, the 2012 European Parliament directive on waste electrical and electronic equipment (WEEE): http://eur-lex.europa.eu/legal-content/FR/TXT/HTML/?uri=CELEX:32012L0019&from=EN Or the new eco-design regulations for household appliances announced in 2019 by the EU: https://ec.europa.eu/commission/presscorner/detail/en/ip_19_5895

7 Fairphone is a social enterprise developing smartphones with a minimal environmental impact. They released a mobile handset free from conflict minerals (gold, tin, tantalum and tungsten), with fair labour conditions for the workforce along the supply chain, and set up to help users extend their phone's lifespan (through modular components designed to be easily repaired and upgraded).

when it has been immersed in liquid, and so on. The services provided vary, but they tend to focus on the material elements of the handset or terminal. Though the problem or issue is usually with the device's hardware, repair technicians may also be able to address software issues; overseeing updates, changing language or accessibility settings, installing applications, or adding software and accessories not supported by manufacturers.

Where phone manufacturers and their authorised subcontractors comprise the formal maintenance system, the latter options encompass a range of unauthorised interventions, set up without the support of telecoms companies, and often skirting the edges of legality. Operating for profit, local repair stores are either established to capitalise on a founder's technical expertise, or else approach phone repair as a way of supplementing another commercial activity—usually something related to technology or telecommunications, whether a cybercafé offering internet access, or a shop selling phone handsets, phone accessories, or e-cigarettes. By contrast, Fab labs, hackerspaces, and other community repair spaces approach maintenance and repair activities as part of a more ideological mission; seeing another way to engage participants in the "opening" and re-appropriation of technical objects.

These secondary and unauthorised repair practices demonstrate similarities with the "maker movement[8]" and the parallel emergence of hackerspaces and Fab labs, as venues for people to participate in—for example—electronics, object design, carpentry, sewing, cooking, and computer programming. Where some commentators, such as Californian entrepreneur Chris Anderson (2012), have hailed these developments as the start of a new industrial model, others, such as Richard Sennett (2008) and Matthew Crawford (2009), see a renewed appreciation for well-established practices of manual and artisanal work, tracking a rising dissatisfaction with consumer capitalism. In extending these interpretations of DIY and "maker" practices to the analysis of smartphone repair, we can see a clear distinction between, respectively, more entrepreneurial and community-oriented approaches to maintenance and repair. Phone repair stores participate fully in the market economy, with technician-entrepreneurs

8 The "maker movement" is a subculture of learning-by-doing linked to influenced by the Do-It-Yourself ideology. Framed in opposition to passive consumer culture, it stresses the importance of turning consumers into producers and engaged citizens. (Anderson, 2012; Dougherty, 2012, Hatch, 2013)

trading their skills and expertise as a paid-for service. Makerspaces and repair cafés set out to empower technology users through the reworking and re-appropriation of technical objects. Usually operating on a non-commercial basis, they demand nothing beyond participants' engagement as part of a community of practice.

These contrasting modes of repair can be further explained with reference to two recent socio-economic shifts. On the one hand, the economy of telecommunications companies is changing, something visible in the slow decline of a business model in which the upfront cost of smartphone handsets was subsidised by telcos, and the resulting growth of a second-hand market for used hardware. A second contributing factor can be identified in a particular social and cultural shift, as a growing subset of users adopt environmentally-friendly behaviour; for instance, taking action to extend the lifespan of their consumer goods. The tensions and overlaps between these two phenomena provide an analytically fertile backdrop for our research. By taking these different repair spaces as our focus, and working to document the practices of those who run and use them, this book offers an insight into how users are resisting and responding to the challenges of technological obsolescence.

REPAIR AND MAINTENANCE STUDIES

From an academic perspective, we see this book as part of an emerging field of repair and maintenance studies.[9] Although failure, crisis and breakdown have long been of interest to social scientists,[10] repair and its preventive corollary, maintenance, have only become a subject of wider theorising in the past 20 years.[11] Writing in 2000, sociologist Christopher Henke made the case for a "sociology of repair," capable of going beyond ethnographic studies of specific repair practices, and addressing the importance of repair as "an integral, though often hidden, skill for the maintenance of social order." (Henke, 2000, p. 75) Further stressing the topic's importance, British researchers Stephen Graham and Nigel Thrift (2007) have explored

9 See Henke (2000), Graham and Thrift (2007), Jackson (2014), Strebel et al. (2019).

10 See, for example, the large body of work on industrial incidents, e.g. Shrivastava (1992) on the Bhopal catastrophe, Vaughan (1996) or Mahler (2009) on the space shuttle accidents, and Plokhy (2018) or Higginbotham (2009) on the Chernobyl disaster.

11 Although ethnographies of repair work abounds, e.g. Harper (1987) on cars and agriculture machines, Orr (1996) on photocopy machines, Moricot (2001) on aircrafts, Dant & Bowles (2003) on cars.

how failure or breakdown can change an object's status—bringing things which are normally ignored back into view. In his 2014 paper "Rethinking repair," Stephen J. Jackson found another reason to focus on these practices; demonstrating how maintenance, repair, and refurbishment provide "a vital source of variation, improvisation and innovation" (Jackson, 2014). We build on this work, describing how repair sites are not constrained by a simple logic of restoration, but participate in the modification, improvement, and transformation of technical objects.

Despite the apparent distance between repair and innovation,[12] a clear-cut opposition of the two is misleading—obscuring the role of creativity and inventiveness in maintenance processes. (Jackson, 2014; Graham and Thrift, 2007) When objects break down, there is a need for new solutions. Small-scale changes made in the course of repair can be a trigger for innovation, "acting as a continuous feedback loop of experimentation which, through many small increments in practical knowledge, can produce large changes" (Graham and Thrift, 2007, p. 5). In his 2007 book The Shock of The Old, historian David Edgerton offers the case of the Japanese bicycle industry, which had its origins in the maintenance and repair of imported British bicycles. Japanese firms started out making replacement parts for these foreign imports, which were then assembled into complete bicycles. Exports of these cheap Japanese-made bicycles grew through the 1920s–30s, to the point where South East Asia was "awash with semi-British, semi-Japanese bicycles, and indeed entirely Japanese copies of British bicycles." (Edgerton, 2007, p. 98) In these examples, the repair and transformative reuse of "old" and "outdated" techniques or objects contributed to innovation processes in ways that are often overlooked.

These processes often begin as "bottom-up innovation" by communities of practice and individual users, who are able to identify their most pressing needs and tailor solutions to their problems. (von Hippel, 1988) Such problem-solving may parallel the forms of creativity, poaching or subversion described by French scholar Michel de Certeau (1980). These creative capacities for

12 "At first glance, nothing could seem farther apart than the apparently separate questions of innovation and repair. Innovation, in the dominant coding, comes first: at the start of the technology chain, in moments of quasi-mythical origination, a creature of garage-turned-corporate engineers, operating with or without the benefits of market research and user experience operations. Repair comes later, when screens and buttons fail, firmware is corrupted, and the iPhone gets shipped back to wherever iPhones come from." (Jackson, 2014) the modification, subversion, and re-appropriation of existing technology comprise a major resource in the development of new products and services. (VON HIPPEL, 1988) For Lucy Suchman (2009), recognising this hidden contribution can shift our analytical focus to places where new products or services are rarely recognised as such. In other words, maintenance work and troubleshooting activities can be seen as a hidden or silent form of innovation, there to be revealed.[13]

From this work, it is clear that practices of repair and maintenance are not constrained to restoring a given object's functionality, or returning it to some original, default state. Exploring the repertoire of such practices, and their possible outcomes, Stephen Graham and Nigel Thrift (2007) invite us to consider:
"the bodged job, which still allows something to continue functioning but probably at a lower level; the up-grade, which allows something to take on new features which keep it contemporary; the cannibalization and recycling of materials, which allows at least one recombined object to carry on, form from the bones of its fellows; or the complete rebuild, which allows something to continue in near pristine condition." (GRAHAM and THRIFT, 2007: p. 6)

What starts as maintenance or repair may become innovation. A user's response to their smartphone's rapid obsolescence can lead to customisation or redefinition. In this, we can see the logic of Jackson's call for researchers to engage in "broken world thinking" (JACKSON, 2014)—taking accidents and technical breakdown as a starting point for new and different kinds of inquiry. Such an approach would draw our attention to objects' social underpinnings, and "the ongoing forms of labor, power, and interest... that underpin the ongoing survival of objects as things in the world." (JACKSON, 2014, p. 230) It would also address the social and material dimensions of maintenance and repair—posing various questions about the composition of such practices. Who takes care of objects and infrastructures? Who responds to technical failures, and how? What skills and knowledge are required to make repairs? As Jackson asks:
"Can breakdown, maintenance, and repair confer special epistemic advantage in our thinking about technology? Can the fixer know and see different things—indeed, different worlds—

13 "A necessary strategy in recognizing those labours is to decentre sites of innovation from singular persons, places and things to multiple acts of everyday activity, including the actions through which only certain actors and associated achievements come into public view." (SUCHMAN, 2009) than the better-known figures of 'designer' or 'user'?" (Jackson, 2014, p. 229)

Engaging with these questions, Jérôme Denis and David Pontille (2015b; 2017) urge us to look beyond the ruptures of failure and breakdown, and engage with the "immense variety" of practices involved in "the maintenance of objects, technologies and infrastructures." (Denis and Pontille, 2017, p. 13) This requires a greater attention to situations and phenomena currently seen as quotidian, marginal, or banal—to the extent they even register as worthy of consideration. Given its near-ubiquitous presence in everyday life, this book's focus on the smartphone offers a point of entry to describe, and explore, the diversity of practices comprising this maintenance regime.

ON PHONES AND SMARTPHONE REPAIR

Work on repair has often taken an ethnographic approach, producing in-depth descriptions of the maintenance of automobiles (Dant and Bowles, 2003), housing (Strebel, 2011), signage (Denis and Pontille, 2015a), and furniture (Gregson et al., 2009). Although research has been done on mobile phone repair, most of this work has focused on the Global South, with representative work in India (Chipchase, 2006), Bangladesh (Jackson et al. 2014), Uganda (Houston, 2014), and Namibia (Jackson et al., 2012). As a result, there has been much less work on phone maintenance and repair in Europe and North America.[14] At the same time, little published work has had to account for the greater complexity of the smartphone, as compared with earlier generations of handsets—something with clear implications for practices of maintenance and repair.

It was this gap in the literature that prompted us to investigate smartphone repair sites in Switzerland. The complexity of these devices, which combine hardware and software with varying levels of openness,[15] justified our focus on smartphone repair. With this as our subject, stores and hackerspaces offered a strategic location to deepen our understanding of the maintenance and repair of digital artefacts. By situating our investigation in Switzerland,

14 Contemporaneous with our project, researchers in the US have studied mobile phone repair shops in Washington DC. (Bell et al., 2018).

15 Depending on the manufacturer, the possibility of changing or modifying these components is more or less limited.

we hoped to demonstrate the limits of a stereotype of the Swiss—and, more generally, Europeans—as quick to buy new devices, and rarely engaging in maintenance or repair. Our choice of setting was also motivated by the relative absence of repair shops and hackerspaces in Swiss public discourse, despite a growing awareness of environmental issues.[16] For us, the mismatch between the relative invisibility of these spaces and their active user base was a puzzle requiring further investigation.

Taking phone maintenance as our subject, and Switzerland as our context, our research asked three main questions:

- What types of breakdown or failure do repairers respond to? What repair practices can meet these needs? And what kinds of repair and maintenance happen in such places?
- How do repairers, often lacking formal training, come to handle the repair and maintenance of these devices? Given that smartphones are complex, opaque artifacts, not intended to be serviced by anyone but the manufacturer, how do they develop these practices and update their knowledge?
- More generally, how should we make sense of what's happening in these spaces? Do these maintenance practices and sites of repair represent a neglected form of innovation?

With these questions in mind, and seeking to address gaps in the existing research, we focused in on those places where repair and maintenance takes place. We visited stores and hackerspaces, hoping to better understand how these practices can improve the durability and lifespan of smartphones.

Throughout this project, our assumption was that an understanding of maintenance and repair is both timely and urgent, a fundamental part of any attempt to meet the challenges of an era of consumer waste and environmental crisis. An inquiry into "mobile repair cultures," presented through a series of portraits of individual repairers, this book may not address these complex problems head-on, but our work can be seen as highlighting laypeople's efforts to improve the sustainability of technical objects.

16 Something we noticed over the course of this project, as growing numbers of colleagues and students, aware of our work on this topic, came to us for store recommendations.

HOW WE DID THIS RESEARCH

This book draws together material from a two-year study of smartphone repairers in Geneva, Lausanne and Zurich. Our field research comprised two phases. We started by conducting a multi-sited ethnography[17] in Zurich and the Lake Geneva region. Investigating repair spaces, we conducted 42 in-situ interviews/observation sessions in independent repair stores around Lake Geneva and 11 in Zurich. Lasting between an hour and a whole work day, these sessions were conducted with both managers and repairers, sometimes interrupted by customers, colleagues, partners, and competitors. These were supplemented with around 60 follow-up visits, and one author's participation in a four-month repair shop internship. In addition, we conducted 13 interviews/observation sessions in hackerspaces and Fab labs in Geneva, Lausanne and Zurich, and made repeat visits to repair cafés hosted by these spaces.

Between 2016 and 2018, we visited most of the stores in Geneva and Lausanne, and those that were the most active in downtown Zurich. The interviews had three parts. First, we asked how they had first come to the business of repair. This was followed by a conversation about the experience of practicing maintenance and making repairs, using customer requests from the day of the interview as a prompt for discussion. Specifically, we asked questions about the diversity of customers' problems, the difficulties these repairers encountered, and the knowledge and skills they developed through their work. The final phase was more observational, as we focused on documenting details of the technicians' spaces, paying particular attention to traces of how they organise their work, their social lives, their tools and access to spare parts, their manual repair gestures, and how they document their practice. We also looked at their reliance on different personal social ties by documenting the media they used in the course of their work (phone calls, videoconferences, WhatsApp groups, posting on web forums). Although most repairers accepted our presence and agreed to answer our questions, only a few had the time to accommodate repeat visits—an obstacle we tried to work around by having one of us participate in an in-store internship.[18]

17 Our focus on multiple locations followed Marcus' notion of multi-sited ethnography (Marcus, 1995).

18 The internship lasted four months and involved helping a manager and her technician in one the shops, participating in everything from welcoming customers to fixing simpler

Relying on photography to document our visits, many of our interviewees expressed reluctance about being recorded in this way, with some explaining that they lacked the necessary work permits, and risked being identified. As a result, we quickly realized that the use of drawings provided a way to put the repairers at ease, especially when starting out; a good format to show the value their work; and a metaphor to quickly explain what we were doing ("documenting your practice to understand cultures of repair"). Making us and our work legible to interviewees, it offered access to other shops, through personal recommendation: "It's Anaïs, the girl who makes drawings of our shop," as one informant told his cousin in Lyon, "you can chat with her." More so than any photos, the drawings proved particularly useful in engaging repairers in discussion, when seeking further explanation of particular details, and in validating our interpretations of what we saw.[19]

We complemented our in-situ interviews and observations with an online ethnography of websites and discussion forums devoted to repair.[20] Our aim was to understand the resources available to beginners and more experienced repairers, and the wider discourse around different user issues, and their likely solutions. Finally, we issued a call for contributions on the research project's website,[21] inviting smartphone users to share their perspectives on any issues they had encountered and how they went about solving them.

Ultimately, we decided to turn the material from our field research into a series of portraits. We selected 16 repairers, identifying a group that best captured the diversity of spaces, locations, demographic profiles, and levels of experience and

hardware problems. It provided an opportunity to spend entire days in the shop, observing not only the repair practices but also the use of the space and the social dynamics at play.

19 For a more extensive discussion of the importance of drawing in ethnography, see Causey, 2017. On presenting research findings in comic or graphic novels see, for example, Galman, 2009; Schwanhäusser, 2016; or Weaver-Hightower, 2017. See also Bosqué, 2015 for another use of drawings in exploring Fab labs.

20 Including iFixit, a wiki-based site which describes itself as 'the free repair manual'; opinions shared on social networks such as Facebook; YouTube video tutorials; debates, discussions and advice on forums; and comments and reviews of particular tools and spare parts.

21 https://head.hesge.ch/mobilerepaircultures/stories/

specialisation we encountered in our research. Given that one of our aims was to investigate how repairers mobilise different resources (formal education, self-instruction, and relationships with their social network) to develop their skills, the portrait format enabled us to represent people's different paths to practising maintenance and repair. It also allowed us to contextualise the different spaces, with reference to their typical customer or user base, the kinds of problems they handled, and their social and cultural environments.[22] Each portrait includes an introductory text describing the store from our perspective; a transcript of the subject's narration of their own route into repair; a selection of photographs depicting the material culture and atmosphere of the space; and 2–3 pages of illustrated ethnographic vignettes. Coming at the end of our field research, we tried to achieve a balance in our choice of vignettes—selecting material that addressed our research questions while also reflecting the diversity of situations we encountered.

Following and reflecting on the portraits, an analytical chapter draws together some of the lessons we took from this field research, as well as the wider implications of the kinds of spaces we encountered, how repairers learn and work as a network, and how repair practices can foster innovation. For readers unfamiliar with the terms used by the profiled repairers, we provide a glossary at the end of the book.

22 We also included two additional portraits from places outside Switzerland. One of them is a wholesale store based in Lyon, opened by one of the repair entrepreneur from Geneva. The other is not a proper portrait, since it is focused on a shopping mall in Shenzhen, where most the smartphone spare parts can be bought. Both offer an important complement to our data.

the Old folk's Repairman

01

THE OLD FOLK'S REPAIRMAN

Two arcades share the same red sign at the entrance of the rue de l'École de Médecine. To the left is an art gallery, on the right is an electronic repair shop, containing bric-a-brac made up of old mobile phone models, radio sets, and all kinds of objects that could be part of the decoration or a side activity undertaken by the clinic's manager.

There is a Ravensburger puzzle, rabbit sculptures, a plastic parrot, a decorative lava lamp, black and white photos of New York, a cardboard cut-out of Valerian, a box filled with pocket books, a film poster for *The Fifth Element*, a Mickey Mouse phone priced at 700 CHF. A sticker warns "no credit given."

Beyond this baroque assembly, behind the "counter," cardboard and plastic boxes of spare parts are skilfully labelled, their contents marked with neat, handwritten stickers. A CCTV camera overlooks a door protected by a curtain. Behind this is the back shop, the actual repair area, filled with machines and tools, cables and chargers hung on the wall.

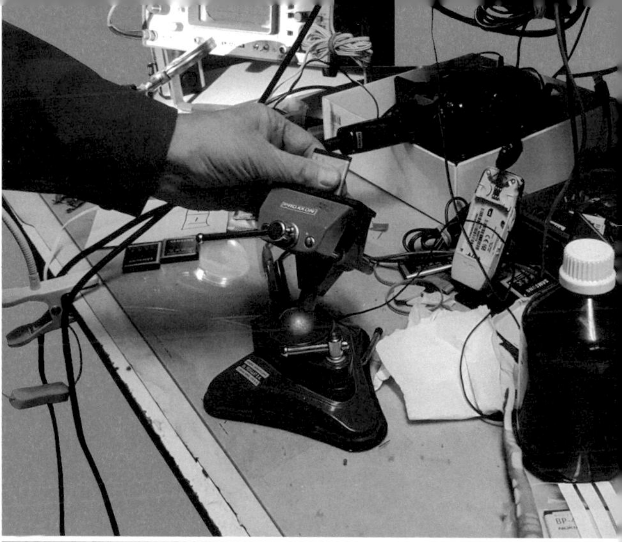
PROXXON

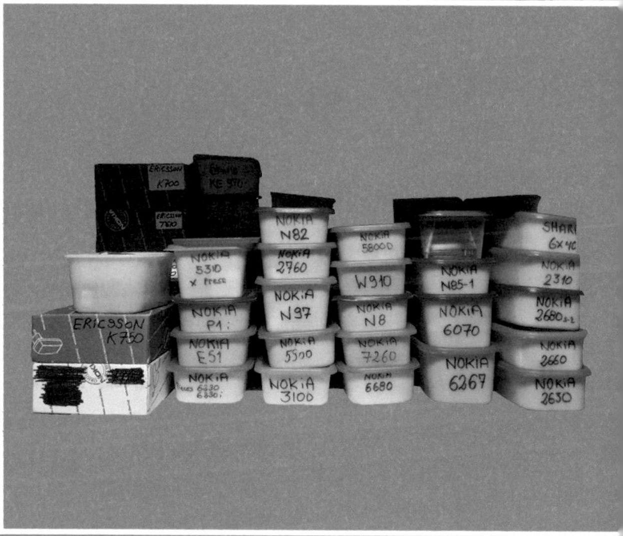
ERICSSON K700
ERICSSON K750
NOKIA N82
NOKIA 2760
NOKIA N97
NOKIA 5500
NOKIA 3100
NOKIA 5310 x Press
NOKIA P1
NOKIA E51
NOKIA 5800D
W910
NOKIA N8
NOKIA 7260
NOKIA 6680
NOKIA N85-1
NOKIA 6070
NOKIA 6267
NOKIA 2310
NOKIA 2660
NOKIA 2630

Article de Décoration

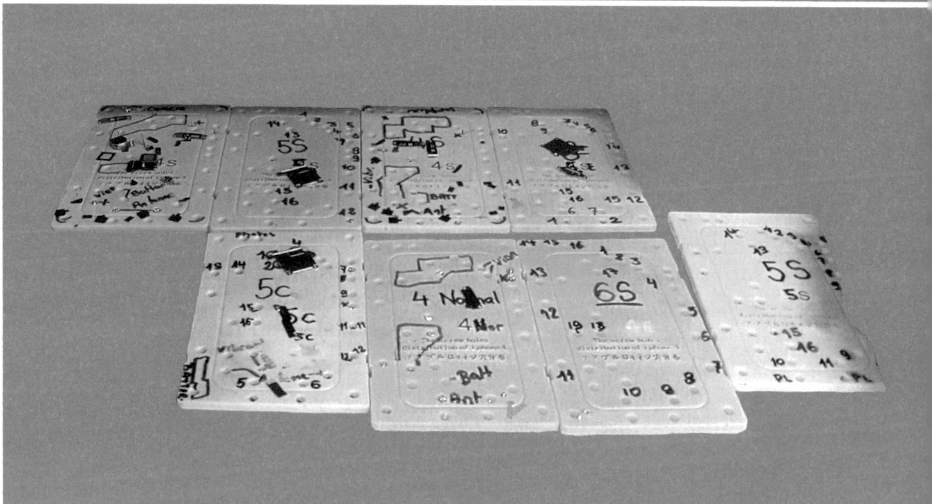
5S
5C
4 Normal
6S
5S

A. (GENEVA): I'm the old folk's repairman! I used to do radio and transmission. I have a background in electronics, then I learned on the job… I started by selling and repairing all transmission systems: radio, television.

At the time, TVs and radios were still being repaired, and we sometimes went to people's homes, you know. In 1980, I started selling and repairing phones. You know, an electronic circuit is always the same. I did some computer troubleshooting too. And I was the first to do so in Geneva. People recommend me, word-of-mouth for thirty years! I even went on a radio show, telling people about the lifespan of phone batteries. They also wrote an article about me in La Tribune de Genève, there are people who come here with the article.

I'm having fun, you know. You have to be inventive to make technology, you have to learn on the job. Every time a new transmission concept appears, we try it out, we try to fix it. I had apprentices, but they were not that motivated. They were interested because they liked phones, but when they saw what it meant, they just wanted to leave. It was too complicated for them, they lost faith. The difference between my shop and the small stores is that I also repair phones that no longer work, while they only replace screens. Today's repair guys are ruining our business. Informal stores only understand new models, they only know how to change screens. It is easy to do that. We used to fix everything! What has changed is the size of the screen, the touch, that's really a big change, the camera, etc. Now we only replace the screen or recover data. Hardware is hardware but the software part affects how the system is programmed.

In the past, telephones were manufactured in many countries, in Hungary and Finland. Now everything is centralized in China. It's difficult to find the spare parts. My advantage is that I have a 30-year-old stock. I'm not throwing anything away! I go to Paris once a month to buy parts from my suppliers.

I'm having fun. I'm 72 years old, I stay in shape, to see the microcomponents I just have to put on my glasses.

A. talks about phones as if they were living things.

LA CLINIQUE DU TÉL
AND THERE'S ALSO THIS GUY WHO WANTED TO RECOVER HIS CONTACTS FROM A NOKIA 6310
I HAD TO TRANSFER THEM MANUALLY, IT TOOK ME TWO WHOLE EVENINGS
I HAD TO PRESS 1, 2, 3, 4 BEFORE MOVING EACH CONTACT FROM ONE PHONE TO THE OTHER !!!

ONE DAY, A MAN CAME TO SEE ME AT THE STORE
I HAVE AN ANECDOTE FOR YOU! IT HAPPENED NOT SO LONG AGO.
HIS FATHER HAS JUST DIED AND HE WANTED TO GET THE PHOTOS FROM HIS PHONE...
IT WAS IMPORTANT TO HIM, THEY HAD SENTIMENTAL VALVES
BUT HERE IS THE THING, THE TELEPHONE DIDN'T TURN ON
IT WAS AN OLD CLAMSHELL PHONE...
SAMSUNG E700
I TRIED TO WORK OUT WHAT WAS WRONG

The story of a guy who wanted to recover photos from an old Samsung phone that had belonged to his father.

Repair Metaphor

I NOTICED THAT THE RIBBON CABLE THAT CONNECTS THE LCD SCREEN TO THE MOTHERBOARD WAS KAPUT!
PHONE WITHOUT ITS FRONT COVER
RIBBON CABLE FOR LCD SCREEN
SO, I TRIED TO FIND A SIMILAR MODEL
FOR ITS RIBBON CABLE
I FOUND LOTS OF OLD MODELS!
BUT THEIR RIBBON CABLES WERE ALL BROKEN
!!!
IT LOOKS LIKE A COMMON DESIGN PROBLEM IN CLAMSHELL PHONES
AFTER COLLECTING 12 MODELS, I FINALLY FOUND ONE THAT WORKED
I COULD RECOVER THE PICTURES
← TUPPERWARE BOX FILLED WITH SAMSUNG E700

tHe WiZARdS

#02

THE WIZARDS

Located next to an organic bulk grocery store, with a similar aesthetic based on unfinished wood furniture, I. is simultaneously a workspace, a computer store and an open workspace. Here and there, posters and flyers highlight the values of openness and commitment. Various devices, often opened and disassembled in full view, suggest everything here is recycled, and reminding us of the importance of being able to open our machines in order to better understand and control them.

While digital objects are ubiquitous—from smartphones to computer screens, plastic robots and tools of all kinds—the default assumption here is not excessive consumption but, instead, a considered, parsimonious adoption of technologies, in a way that's more virtuous, and with a smaller environmental impact.

ITOPIE
SOCIÉTÉ COOPERATIVE
SOLIDAIRE
Pour une informatique OUVERTE, LIBRE, respectueuse des personnes et de l'ENVIRONNEMENT
le petit conseil est gratuit
OFFRES AUX PERSONNES
- Vente matériel d'occasion professionnel et garanti
- Sensibilisation et introduction au libre
- Réparations
- Accompagnement technologique
d'une alternative
OFFRES
- Sauvegarde données
- Installation GNU linux + 1er cours
- Changement écran
70.-
50/120.-
itopie

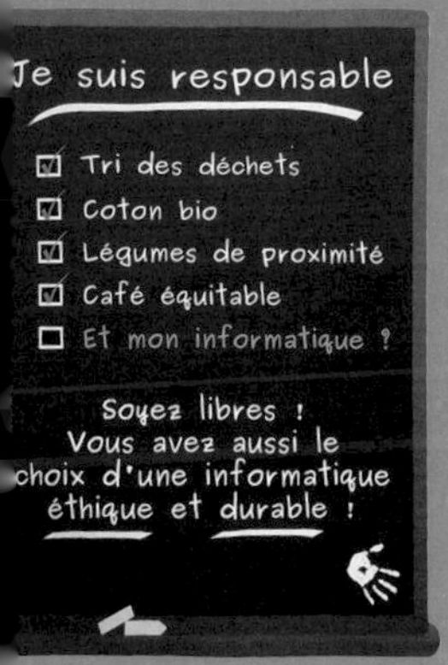
Je suis responsable
☑ Tri des déchets
☑ Coton bio
☑ Légumes de proximité
☑ Café équitable
☐ Et mon informatique ?
Soyez libres !
Vous avez aussi le choix d'une informatique éthique et durable !

FAIRPHONE

2000

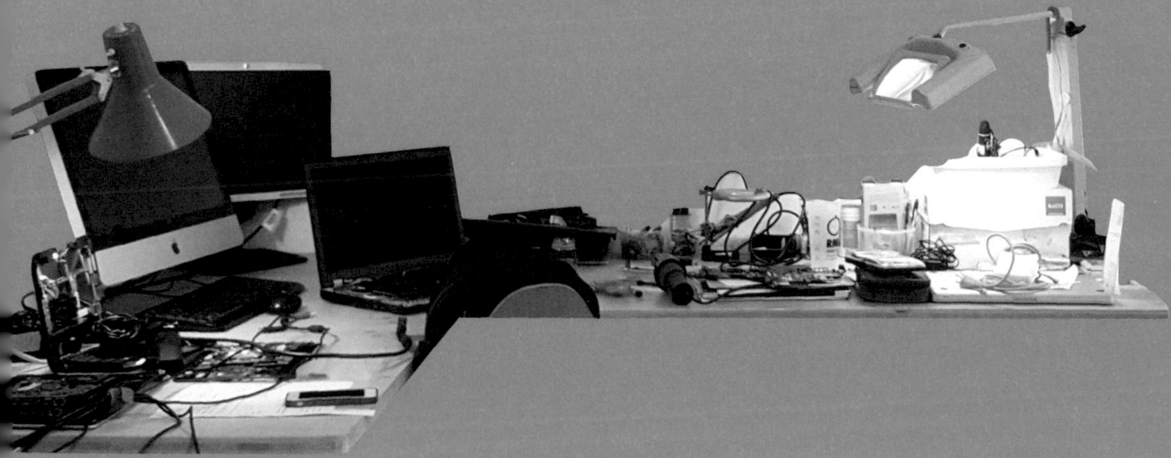

S., (GENEVA): I worked for 16 years in a luxury business, in their IT department. After a while, I no longer agreed with the company's philosophy and so I decided to join a non-profit cooperative fighting for free, ethical, sustainable and socially responsible IT. This project started in 2009 but the cooperative really took off in 2012. We're now 42 members, including three people who are both co-owners and employees. I am in charge of outsourcing, E. who was a developer in Argentina, and J., the IT system administrator.

Overall, we have two main activities. Firstly, activist computer work, collaborating with organizations or companies such as a local coop who is already aware of these issues and who knows us: we set up their servers, their networks, Wifi, etc. We have about fifteen customers, and we try to provide ethical solutions, avoiding solutions that go exclusively through the cloud, favouring free software, promoting hardware such as the Fairphone. The second activity involves fixing things. We repair digital equipment and sell things that we have refurbished, with a one-year warranty. E. is in charge of the repair and second-hand sales department. He also has a workshop project where participants disassemble computer hardware to extract and recover the components.

J., (GENEVA): When it comes to repairs, it's more about computers than smartphones. But we sometimes advise customers, help them diagnose the problem, and then send them to nearby stores. There are many little shops that can take care of this around here in downtown Geneva. We know them.

Also, the cooperative has recently been opened to members who are, in a way, non-experts, "sympathetic" to our cause. What is interesting is that we have since had more and more women participating in our activities. Currently, we even have a majority of women members. One of the members—an intern trained in open-source communication tools—organizes crypto-parties.

We also have evenings in which the wizards—the computer experts— accompany the "Muggle," the non-experts, a Harry Potter reference. This is a technocratic world, and people who do not have the right tools are being stepped on, so it's important to show them. We are currently looking for funds to improve our workshop and buy machines that will allow us to grow. The problem is that the cost of labour in Geneva is very high.

libre
"WHAT'S INTERESTING IS THAT THERE ARE MORE MORE WOMEN IN OUR WORKSHOPS"
"IN FACT, MOST OF OUR MEMBERS ARE WOMEN"
énergie grise

"One of our members organises crypto parties evenings when 'wizards' computer experts help 'muggles' (non experts) to install open-source software…"

THERE ARE ALSO BIG DIFFERENCES BETWEEN BRANDS AND THE QUALITY OF THEIR PARTS
FAIRPHONE
FAIRPHONE
OPENED IPHONE ON THE TABLE
THERE ARE DEVICES MADE BY PROFESSIONALS... WE CAN SEE IT RIGHT AWAY YOU CAN SEE THE DIFFERENCE IN THE DESIGN. SOME PARTS ARE MORE SENSITIVE TO HEAT, SOME ARE ODDLY PLACED, FOR EXAMPLE, NEXT TO THE PROCESSOR.
MACS HAVE HIGH QUALITY COMPONENTS, BUT THE SOFTWARE CAN BE TRICKY. SOME MACHINES DON'T EVEN HAVE COOLING FANS ON THE BOARDS WHERE EVERYTHING IS SOLDERED.

THE OTHER DAY A WOMAN SHOWED UP WITH HER COMPUTER
SHE WAS SCARED BECAUSE A POP-UP MESSAGE
HAD ASKED HER TO UPDATE THE SYSTEM.
SHE DIDN'T KNOW WHETHER OR NOT SHE SHOULD ACCEPT...
A 70 YEAR OLD WOMAN! SHE WANTED TO SECURE EVERYTHING ON HER COMPUTER... SO WE EXPLAINED THE ISSUE... AND NOW SHE'S USING PROTON MAIL!

I'm Like a therapist

#03
"I'M LIKE A THERAPIST"

Walking through the store door, in a space much less overcrowded than the other shops we visited, we realize that nothing is out of place, except for a human-sized model of a 3210 Nokia handset to the left of the entrance.

While the walls offer different phone casings and accessories, the overall atmosphere is a little more sober than usual. Behind the counter, a half-drawn curtain reveals a small repair workshop, well guarded by its owner.

iPhone 4 New
iPhone 5s New
iPhone 6s New
iPhone 4/4s occasion
iPhone 4 New
iPhone 5s New
iPhone 6s New
iPhone 5 occasion
iPhone 4s New
iPhone SE New
iPhone 6s+ New
iPhone 5c occasion
iPhone 4s New
iPhone SE New
iPhone 6s+ New
iPhone 5s/SE occasion
iPhone 5 New
iPhone 6 New
iPhone 7 New
iPhone 6 occasion
iPhone 5 New
iPhone 6 New
iPhone 7 New
iPhone 6+ occasion
iPhone 5c New
iPhone 6+ New
iPhone 7+ New
iPhone 6s occasion
New
iPhone 6+ New
iPhone 7+ New
iPhone 6s+ occasion

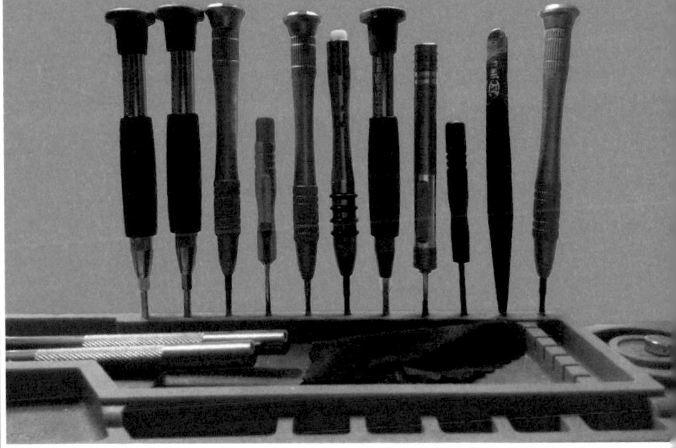

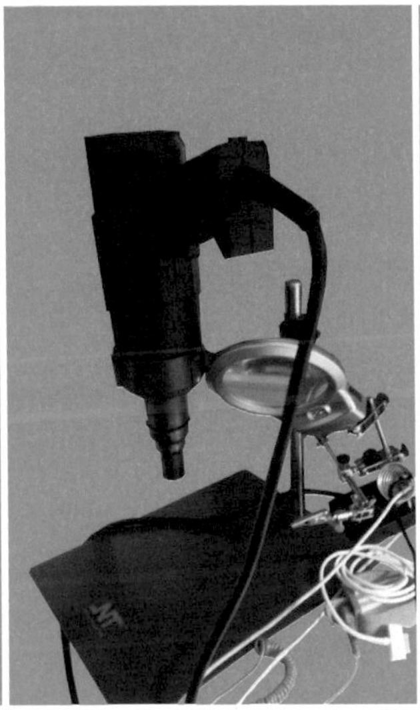

LED OK

Fidget
Spinner
CHOOSE YOUR STYLE
Coated steel ball bearings
for long lasting spin
Helpful for deep thinking and
stay focused

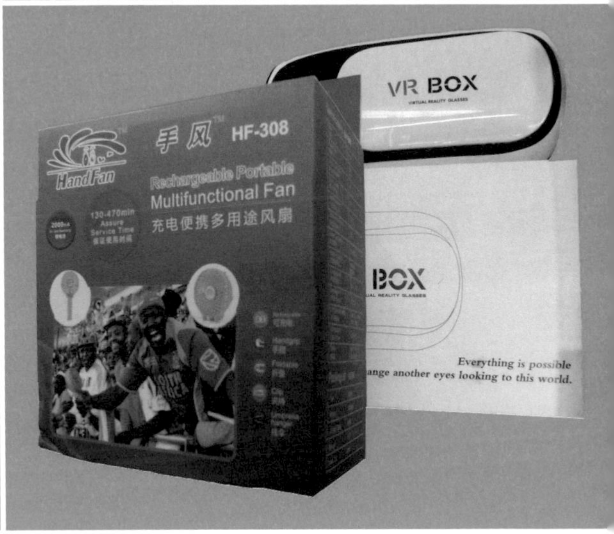
VR BOX
手风 HF-308
HandFan
Rechargeable Portable
Multifunctional Fan
充电便携多用途风扇
BOX
Everything is possible
ange another eyes looking to this world.

S., (LAUSANNE): I am a 26-year-old Swiss woman of Kurdish origin. We opened the store in 2016 while I was still a student. I knew someone who had several stores and it was too much for him, so he offered me the opportunity to buy one. I didn't accept right away, it took me almost three months to assess the risks. I didn't want to do a PhD after my master's degree in political science at the university, and my husband was already working in the mobile telephony sector, selling phones. I thought it would be a good opportunity to open a repair shop. I was lucky that my family could support me.

For me I think it's more difficult because I'm a woman, and especially because I'm young. It's hard to deal with people who don't always take me seriously. I have a foreign-sounding name and I'm a woman. Some customers think I can't fix their phones. Some people ask me, "Did you learn how to do that, or?" But when I tell them the guarantee is six months, when I give them their phone back, it definitely reassures them. There is a lot of questioning of our professionalism. It is a matter of credibility with employees as well, you know. They had to adapt to the fact that their boss is a woman, and younger than they are.

We have created a family-like relationship with clients here, but we must be careful to keep a certain distance in order to stay professional. We become their therapist, we listen to them, we console them… They listen to our advice and offer us gifts and flowers as thanks. I think my feminine sensibility helps too. If the store is clean, it projects a better image and helps customers to have confidence in the business. Basically, there is a link between store maintenance and phone maintenance! This is also why I have introduced a dress code, I want my employees to be well dressed to give a good impression. As for their skills, some of them have been trained elsewhere, for example, in France and Spain, in countries where repair is more mainstream. The job of telephone repairer arrived in Geneva first because it is close to France. On our side, we plan to expand here, and in other large towns in the German-speaking part of Switzerland. We now have several technicians, some do micro-soldering, others, who learned on their own, do simpler things.

After opening this store, I opened another store in Bex. The big difference here is that it's a village and so people are more suspicious. If they don't know about ut, they won't come. The advantage for me is that my mother has a hair salon there, so she recommends my place; it works a lot by word of mouth. I'm lucky that my mother knows the whole village, and if she does a good job, people think I do too! Our business actually works a bit like a hair salon. Once you find a good hairdresser, you always return to the same place! Besides, there's no competition in this town."

RÉPARATION EXPRESS
36
ONE DAY, A MAN THOUGHT HE COULDNT USE BLUETOOTH
BECAUSE HE HAD CHANGED THE SCREEN HERE THE WEEK BEFORE

I HAD TO DISMANTLE THE WHOLE PHONE IN FRONT OF HIM
TO SHOW HIM THAT WAS NOT THE ISSUE
Paramètres

WE ALSO HAVE DISHONEST CUSTOMERS WHO ARE ALWAYS HAGGLING OVER PRICES
WHO ACCUSE US OF WRONGDOING
FOUL-MOUTHED PEOPLE
THE WORST WAS A UNIVERSITY PROFESSOR! HE SHOWED UP IN PANIC HIS SCREEN WAS CRACKED AND HE NEEDED HIS PHONE TO GIVE HIS CLASS
THAT DAY, WE HAD NO SCREENS FOR HIS PHONE MODEL IN STOCK. WE TOLD HIM WE COULD TEMPORARILY INSTALL A FAULTY DISPLAY ...
WE EXPLAINED CLEARLY

AFTER THE WEEK-END, HE CAME BACK SUPER AGGRESSIVE
HE HAD BEEN TO ANOTHER SHOP WHICH SAID HIS SCREEN WAS DEFECTIVE
FORTUNATELY, C.J. HAD EVERYTHING ON HIS CUSTOMER FILE...
THE MAN WAS SHOUTING: " YOU BUY PHONES FOR 3.– "
I HAD TO ASK HIM TO LEAVE

IN BEX, SOME OF THE COSTOMER ARE A BIT OLDER
ONE DAY, AN OLD WOMAN CAME IN SHE FELT LONELY, AND FAR FROM HER FAMILY
IN ONE WEEK, I INSTALLED EVERYTHING FOR HER SHE NEEDED TO SKYPE, SEND AND RECEIVE PHOTOS
SINCE THEN, SHE HUGS ME WHENEVER WE MEET

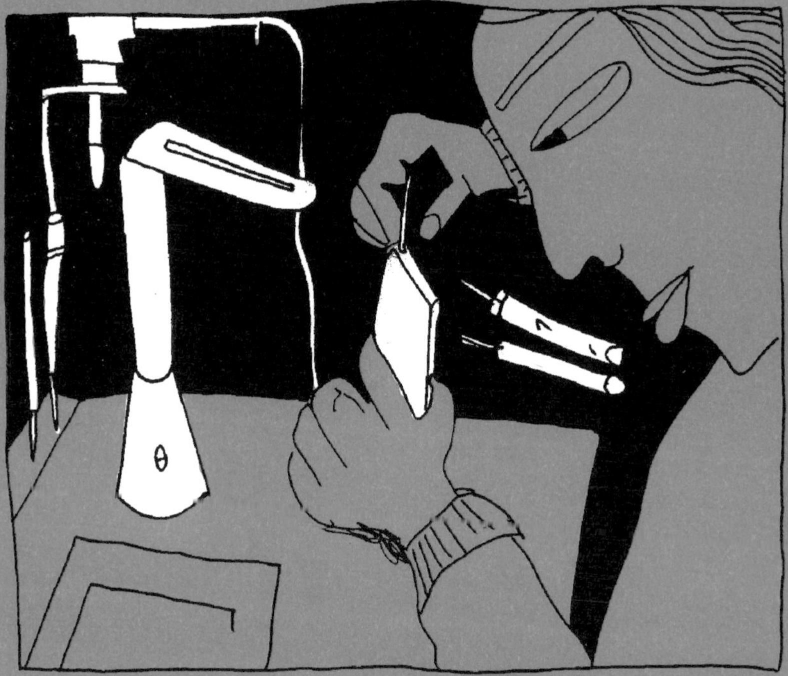

Master of Micro-Soldering

#04

MASTER OF MICRO-SOLDERING

We first met A. in Lausanne, before he disappeared for a few months. More recently, we saw him reappear in different shops in Geneva, where he was helping several of his contacts. Four store owners told us about him and his skills.

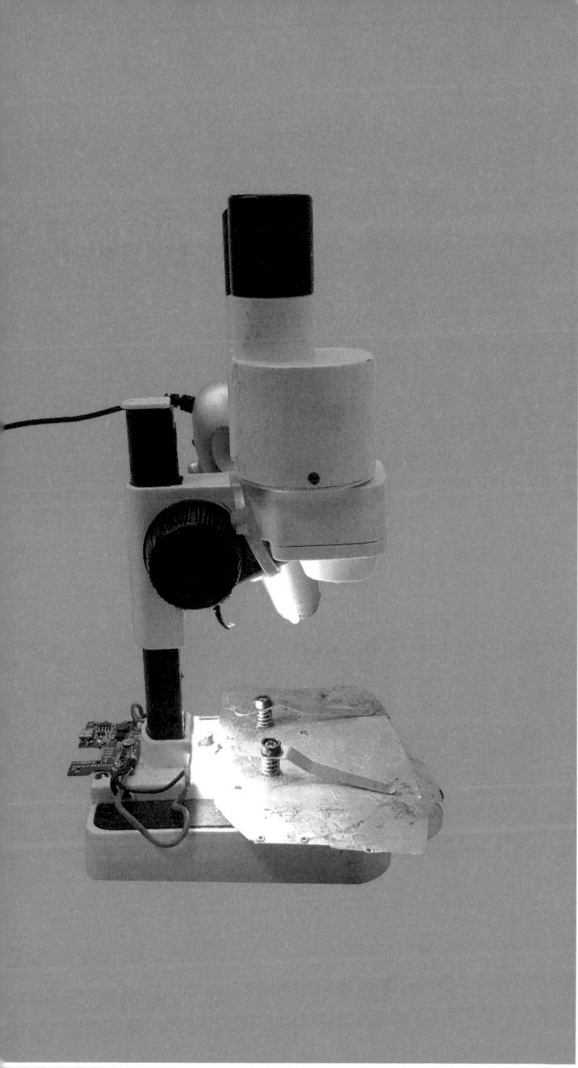

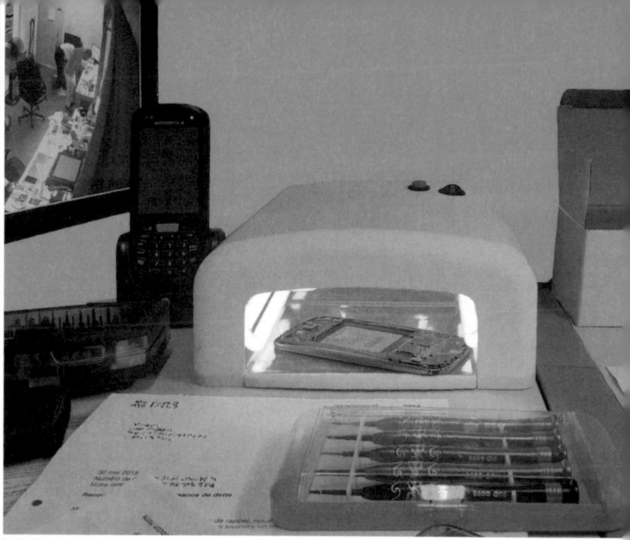

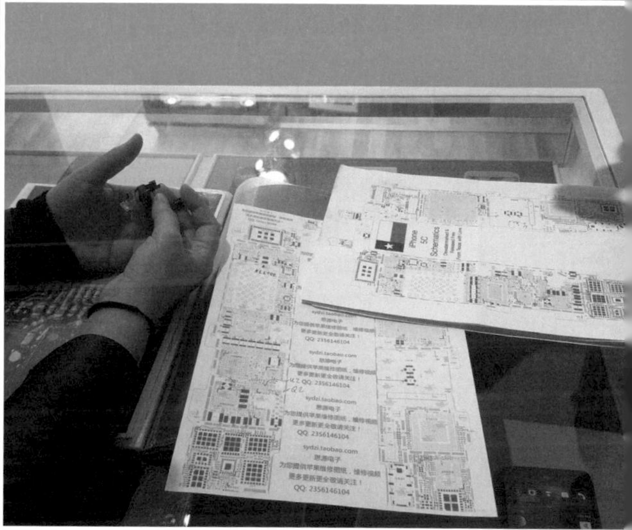

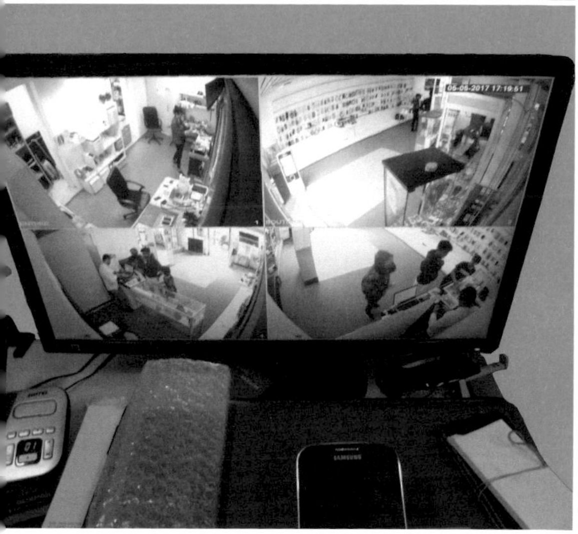

A., (INDEPENDENT): My cousin worked at a store in Geneva, he was the one who brought me to Switzerland. I am Algerian, but I lived in Spain where I started repairing phones, then I lived in France, in Lyon, where I worked in one of the many stores at La Guillotière. I was the master of micro-soldering! I do more complex repairs, not just broken screens, and that's why I have a good network. We all know each other, in Geneva, Lausanne, Thonon les Bains, Lyon.

Now I'm independent, I help the two shops in this neighborhood.

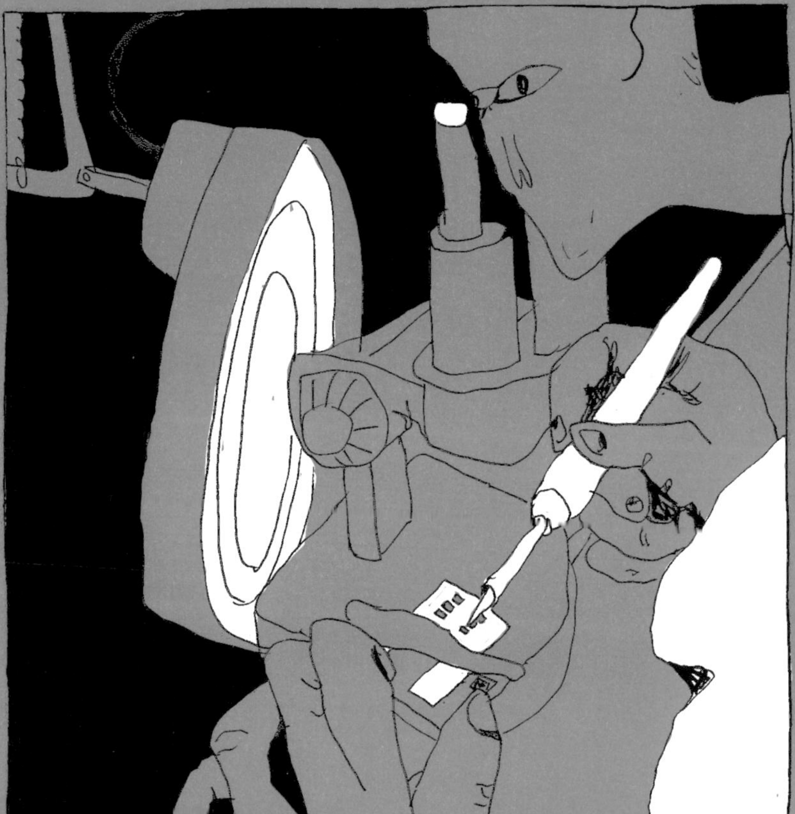

An Eritrean woman showed up.
She wanted us to send an e-mail (for her friend)
to confirm a doctor's appointment next week.

HELLO, WE HAVE TWO BROKEN PHONES... WE CAN'T CHARGE THEM
MINE IS AN iPHONE, HiS iS A SAMSUNG
IT MAY COST TOO MUCH TO REPAIR BOTH OF THEM
CAN WE FIND A SOLUTION?
LiKE... DO YOU HAVE A CHARGER THAT WORKS ON EVERY MODEL?
IS THERE SUCH A THiNG?

A FRIEND
A. BUSY REPAIRING

A NEIGHBOR SHOWS UP AND SAY HI

S., the owner, documents how A. repairs a broken phone.

THE RE-CONDITIONING LAB

#05
THE RECONDITIONING LAB

In a busy street, between a gym and a convenience store, this store has a sign with the name Telecom Corner, whose choice of colour and typography seem inspired by the international photo chain YellowKorner.

Unlike most repair shops, the stalls inside have little in the way of telephone accessories, as if the shop specializes in a different niche. The abundance of big machines and containers for tools and spare parts catch our attention.

Devices from less familiar brands, such as NGT, fix, TBK, or SuQian Xinghe Engineering, have interfaces in English and Chinese. The design and appearance of these devices reveals little of their purpose. We're not entirely sure this shop is meant for regular users.

Votre 1er laboratoire
I 6
I 6+
I 7
I 7+

NGT SOLUTIONS

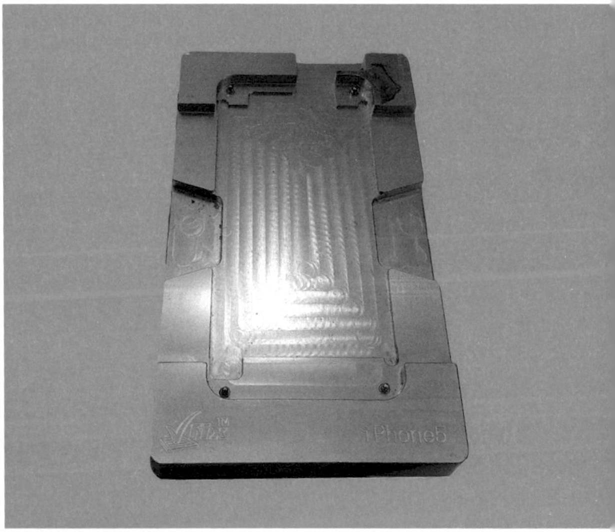
iPhone5

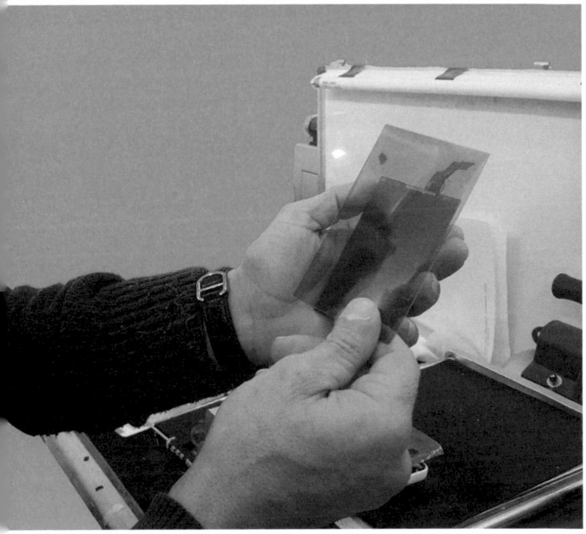

H., (GENEVA): I was born in Morocco, to a Moroccan mother, and a Sicilian father. He was a salesman, selling cars and Arabian horses. As a result, I have made many trips back-and-forth between Morocco, France and Spain. And then I met my wife, whom I married, and we came to Geneva because her sister had an apartment here.

Basically, I started working in telephony to sell accessories and phones, but I always had an interest in repair. I met a guy here who sold jewelry and fixed phones. We collaborated. I really think collaboration is important, and in telephony there are many things to do… The difficulty in this business is the people you work with. For example, I tried to collaborate with a jeweller here in Geneva. I wanted to customize smartphones, put gemstones in them, make a gold chassis for iPhone 5–6. I have a lot of ideas. These are things I did in Morocco where I had a workshop with my cousin, who's a welder. Morocco is a platform for developing Africa, it's full of opportunities, that's what Jacques Attali says! With my cousin, we used to customize phones together, we made accessories, mainly in gold and silver. I met people to try and understand their skills, how moulds are made, we even made a forge with a kitchen mixer. In general, I am fascinated by machines, I want to see and understand. Then I went to Dubai, that's where you really understand what's going on. The Saudis, they're lost in the desert, they buy parts and make custom phones for their rich customers. Here it's different, the guy who makes kebabs sees this and changes jobs, he becomes a repairman. There are real investors in Dubai.

Repair and spare parts have become a niche for me. Sales is a humiliating job, though. I had set up a innovative sales system in 2014. I was the first to do that, I put one at the station, and then in the kiosks in Geneva. At that time, there were offers where you could buy fifty phones at Interdiscount or at the market, I bought everything, and I sold the second hand phones one by one for 10 or 20 francs, making a small profit. I was travelling around among my friends, kiosks and freelancers to offer this service. But it was exhausting.

The store here, this arcade, it belongs to another guy. I offered him a collaboration, this reconditioning laboratory, especially for used screens and displays. In fact, we have plans to open a lab in another store by January or February. I set everything up myself, I even wrote a book about the whole process. I bought the machines in Thailand and brought them back here.

THE RECONDITIONING SPECIALIST
I HAVE TO INTRODUCE YOU TO SOMEONE. HE'S THE MASTER OF MICROSOLDERING.

Smartphone steel mold for LCD display-phone 5.

LCD gel separator machine for curved displays

we are like a family here

#06

"We are like a family here"

We arrive just before opening hours, and are welcomed by yellow graffiti on the store's metal shutters—a large comprising letters and a blue phone, on the screen of which we can see the name of the business.

Once the store has opened, our attention is drawn to a LED panel, displaying a scrolling message that instructs us to like their Facebook page. Behind the counter, a young employee wears a sweatshirt emblazoned with the message "We came for the Future." In front of him, the store's waiting area is empty. Phone casings hung from the wall face a window full of screens, smartphones, and tablets. In the back room, which is off limits to customers, the repair area includes of a desk, a set of tools cleverly stored to one side, and a vast drawer cabinet filled with spare parts. Nothing is out of place, everything is neatly organized.

DEBLOCAGE
de tous vos téléphones
KING OF GSM
REPARATION
Express
tous mobiles,
tablettes &
ordinateurs.
DEBLOCAGE &
PROGRAMMATION
TOUS MOBILES
iPhone
BlackBerry
NOKIA
SAMSUNG
htc
smart mobility
LG

ECRAN LCD

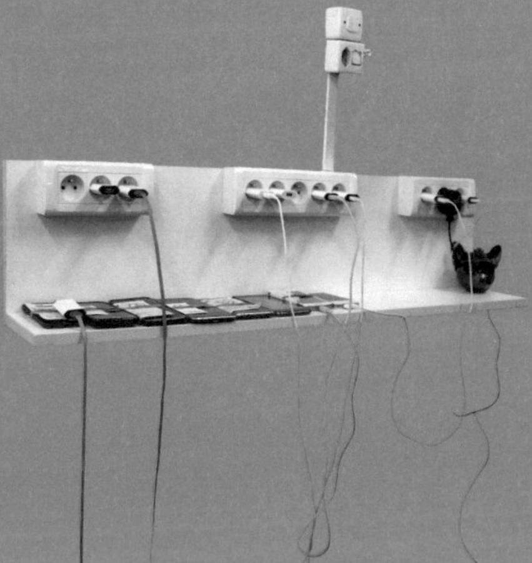

KING OF GSM
SPÉCIALISTE SAV TÉLÉPHONE MOBILE, TABLETTES ET ORDINATEURS
Code
Le 20
BON POUR : N° 0186
Réparation
Désimlockage
Flash
Marque / Modèle
Description
Nom
Tél.
Rue de Berne 58, 1201 Genève - Tél. 022 732 71 70
kingofgsm.ch - E-mail : kingofgsm@hotmail.com
Payé
Tarif : 300.-

B., (GENEVA): The first store I opened was in Annemasse, a town on the Swiss border, near Geneva. I had just gradudated from a French university, having studied marketing and sales. It was 2002, and I had started working as an insurer in a bank in Haute-Savoie, but I didn't like the job. My father comes from Lyon, he was originally from Algeria. We lived in a village in Haute-Savoie and I studied in Bonneville. Before that, I was a basketball referee, coach, player and club president! I even started a civic association to bring medicines to Niger. But I didn't like the insurance business, so I found myself selling phones in Annemasse.

That was when I discovered this passion. Why? Because there was a human relationship. In the store, subscriptions were sold to customers. From the beginning, I saw there was a niche to explore... I took time to explain to customers what was wrong, how best to use their phones, and I realized that there was a need for repair services. Three in five customers asked me where they could fix their equipment. That was in 2003. At the time, I sent them to Lyon, which is like 150 kilometers from here! Three months later, I opened my shop. After a year and a half, I realized that 40% of our customers were Swiss, so I thought we could open there. In Geneva, there was nobody yet, except HED Diffusion, but his business was in decline, so he often sent me his clients.

I opened my first store in the Plainpalais area, close to the university. It was the biggest mistake of my life. I thought that students were the biggest phone users, and there are a lot of them in the area... but this was not the right clientele, because students are broke. Then I realized the biggest phone users are the diaspora! They use their phones a lot because it is the only link they have with their families. It was this population I had to reach... The people who made the business work came from Les Paquis, a very international neihghborhood. I opened here in 2007, and the business took off! Most people who come from the Third World need self-esteem, they need to value themselves. As we started to grow, we opened three more stores, covering each corner of the city.

We have reached a point where they are no longer customers, they have become friends, we are on familiar terms, we are with family here. Everyone appreciates this proximity. Poor people, bobos, ambassadors who come to change their phones... this proximity can't be found anywhere else. Before, in the service business, we used to find this with the hairdresser and the baker. Even today, at the butcher shop, you have to take a ticket to get in line. This is not the case here, we're close to our customers.

Sometimes, the shop is like a social club, with repairers from other stores.

S. helps a customer fill-in his tax declaration.

80% of our sales are made on the wholesale market

#07

"80% OF OUR SALES ARE MADE ON THE WHOLESALE MARKET"

In the Guillotière district of Lyon, close to Cours Gambetta, and opposite an Asian supermarket, there is a small shop much more discreet than those larger repair stores a few dozen metres above.

A billboard promises "repair equipment, spare parts, and accessories for phones, tablets, and ipods for professionals & individuals." The schedule is generous. The shop is sometimes open until 10pm on weekdays, and by appointment on Saturday and Sunday. On Tuesdays and Thursdays "coffee is offered until 10am," something indicated by a small sign at the entrance. The walls are covered with pouches, casings, and screen protectors, while the stalls are filled with pliers, soldering irons, and tubes of all kinds. Behind the counter, a large piece of furniture seems to hold the treasures that customers and couriers come to find, its drawers covered by a French flag.

NBA

SELECTED PROFESSIONAL TOOL
15€
New

B., (GENEVA): I had become my own buying centre, in Annemasse and Geneva, and I thought, why not sell the supplies to other stores? So I opened the store in Lyon. Why Lyon? And why this district? Simply because there are more than 35 stores spread over 200 square meters there. With the idea of opening a regular shop in Lyon was in the works, I decided to take advantage of this synergy without being seen as a competitor to the others.

A., (LYON): Our main business, I would say 80%, comes from the sale of spare parts to professionals and individuals, and this means selling batteries, screens, charge connectors, loudspeaker, connectors, mats, and so on. We have all the tools for professionals, but we sell them to individuals as well. Pliers for everything that needs handling, glue for screens. This is the ultimate kit, with all the tools and small compartments. It costs ten euros. We also do some repairs. I don't do them, I subcontract them, because I don't have a repairman here, so it's a side-activity. We are supporting the Geneva store which makes 90% of repairs. This is B.'s idea. He's spent years travelling around China, meeting suppliers and working to create a complete chain, from spare part sales to repair.

We worked on a kind of organization, established by B., with small lockers so that everyone, you and me included, can find what they are looking for. For example, let me look on my iPhone 5 computer tool, which redirects me to the drawer that corresponds to what I'm looking for. So, I will have access to the LCD screens, the charger, the charging connectors, and so on. In our business, we are constantly adapting, we have to be the first to react, with other furniture we put in place—here, here or there—knowing that we will remove parts and references from our digital catalogue. It's quite a difficult job this referencing and organization, but it is also strategic work, in that we need to choose the right parts. Sometimes we choose parts and we'll realize that it's a flop, that it doesn't sell well, so it's complicated... We have to keep up with the displays, so we work closely with S., for example, who tells me, "well, I have a reference that keeps coming back, it's the iPhone 7, the loudspeaker parts, there's a recurring breakdown." That's why we work as a team, we need to know what to buy and include in the database.

I won't hide that it's a lot of work! The average customer thinks we just sell phone cases, he doesn't realize the complexity... because the telephony market is constantly evolving, according to the different demands and the devices on the market. Samsung, for example, is developing phone models to compete with Apple, which, on on their side, are increasingly restricting phones. So we have to find solutions to bring to our customers... it is a daily headache.

I came from the world of car sales and advertising. I started in repair at the end of September 2014, and by December, I was already operational. We have to be responsive to customers' needs. They have no time to waste, because it's their meal-ticket that needs repairing! Not all the stores you see up there can have parts in stock. This was a strategic decision by B.. The business model is basic, "you go down the stairs and you have what you need in our shop." It's an additional expense to order in China or elsewhere, you have to look, is this the right reference? Now you know that you can go down the stairs and have professionals to welcome you.

OFFICE
STOCK OF SPARE PHONE PARTS
STOCK
PRIVÉ
PRIVÉ
FEAR
ASVEL
BIG CALCULATOR

"I REMEMBER B. USED TO ASK ME TO WRITE OUR NAMES WITH BIG LETTERS ON ALL OUR PARCELS. I THOUGHT IT WOULD BE LIKE TAKING A PACKAGE TO THE POST OFFICE... BUT IT'S VERY DIFFERENT THERE! IT'S 1000 OR 10'000 TIMES BIGGER! THEY HAVE NO TIME, THEY HAVE NO TIME AT ALL!"

save the planet

#08
"SAVE THE PLANET"

In the brand new district of Europaallee, to the side of Zurich's main station, the store catches the eye with its stands made from recycled wooden pallets, green carpets, and precise arrangement of second-hand electronic items.

At the foot of a building with a sleek, modern look, the shop's minimalist decoration and neat interior layout reveal an attention to detail that is more reminiscent of photographs of hipster shops than the colourful storefronts of Langstrasse, a few metres away.

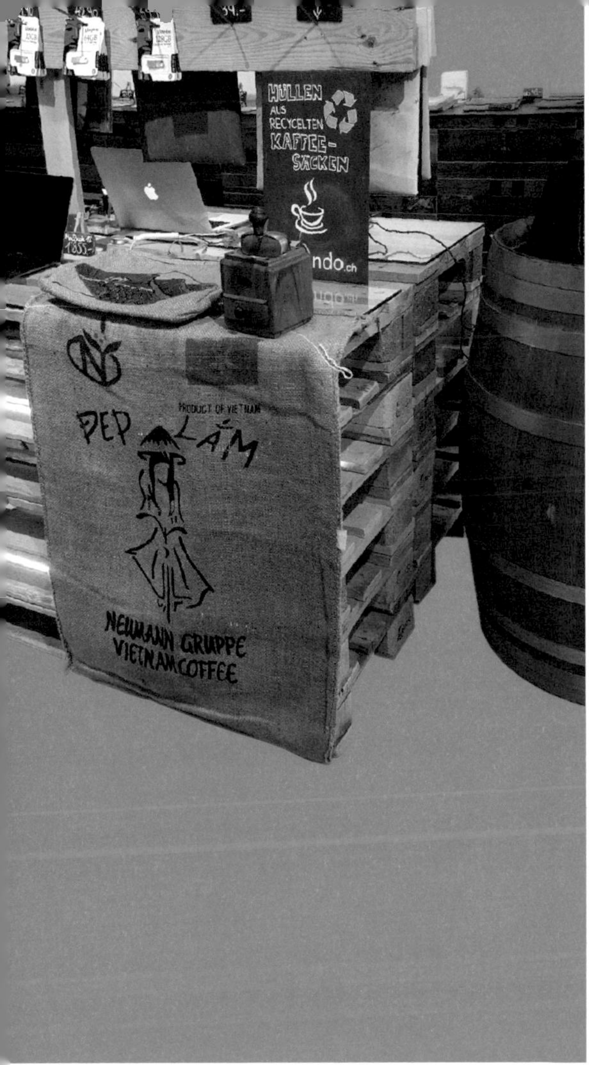
HÜLLEN
ALS
RECYCELTEN
KAFFEE-
SÄCKEN
PRODUCT OF VIETNAM
NEUMANN GRUPPE
VIETNAM COFFEE

Mein
iPad
Mein
iPhone
für iPhone 6 und 6 Plus

IST IHR APPLE
GERÄT DEFEKT
ODER LANGSAM?
KEIN PROBLEM!
GERNE BERATEN WIR SIE
IM SHOP!
IHR
revendo.ch
TEAM

L., (ZURICH): The owners of the store are two guys from Basel who came up with the idea of opening a computer reconditioning service. When they were finishing their studies, they did not want to work for a company. They started buying back machines that they were refurbishing, then selling them online. Why buy new when you can reuse renovated equipment? It was an obvious path for them. Here we have an environmentally friendly approach. We try to use parts from other equipment and recycle everything.

At first, the founders worked from home, but as the demand was soon pretty high, they needed space to store their equipment. So they joined a shared workspace by renting half of the space to other contractors. Then they realized that it was, again, too small and that customers first wanted to see what they were going to buy. So they opened a store in Basel with four employees, and that's where they started to develop the repair business. Since 2015, they have opened four other stores, and I have been in this one since the beginning. I started as an intern, and I learned little by little, by watching videos on YouTube. We also have an internal documentation system where we share information with other members of the store.

Today, there are fifty of us in this chain. There are many roles—managers, salespeople, professional technicians in Basel with their Apple certificate and the right tools, people who manage the hotline or web design, trainees and specialists in logistics and IT.

We don't want to work with Apple because we don't share the same values. Here we are against the consumerist system, we are trying to make second-hand electronic goods, which are better for the economy. Our goal is to save the planet.

We just do quick little repairs in this shop. Five of us work full-time on computers, tablets and smartphones. When we can't fix a device, we send the phone to our Basel centre, or we tell the customer to go to Handydoctor on Banhofstrasse, we know they're good. The Basel HQ has more space and technicians who can spend time with more complex operations and work on the software.

WHAT WE DO WITH THE BROKEN SCREENS?
ACTUALLY, WE WORK WITH
SOMEONE WHO COME ONCE A MONTH
AND COLLECT THEM
HE COMES FROM ROMAMA
SOMETIMES THE DISPLAY STILL WORKS EVEN THOUGH THE GLASS IS BROKEN
HE THEN PEELS OF THE SCREEN
TO REPAIR THE DISPLAY
THIS PROCESS IS CALLED RECONDITIONING
BUT
HERE, THE WORKFORCE IS MUCH MORE EXPENSIVE
SO WE DON'T DO IT HERE
SO IT IS GOOD FOR HIM AND FOR US

YOU KNOW, IT BECAME MUCH HARDER TO REPAIR
BECAUSE THE WAY APPLE PRODUCES TH DEVICES
THE PARTS ARE MUCH MORE EXPENSIVE. APPLE OFTEN MAKES TRICKS TO MAKE IT MORE DIFFICULT TO FIX. FOR EXAMPLE, THEY DON'T SELL DISPLAYS FOR REPLACEMENTS. SO WE BUY DISPLAY WITH NO APPLE CERTIFI-CATE WHICH ONLY WORKS ON THE LOWEST BRIGHTNESS!

WE ALSO HAVE AN INTERNAL LEARNING SYSTEM AT THE HEADQUARTER IN BASEL BUT WE DON'T GIVE CERTIFICATE
SBB
SBB

AND THE LENGHT OF THE COURSE DEPENDS ON YOUR SKILLS!
UBS
repair center
repair center

WHEN WE CANNOT REPAIR DIRECTLY, WE EITHER SEND THE PHONE TO THE HEADQUARTER OR WE TELL THE CUSTOMER TO GO TO HANDYDOCTOR IN BANHOFSTRASSE. I KNOW THEY ARE GOOD. I HEAR GOOD THINGS ABOUT THEM FROM OUR CUSTOMERS!
IPHONE 4/8
IPHONE 5/5S/6+

MY PHONE
MY STYLE

#09

"MY PHONE, MY STYLE"

Though most repair shops also sell smartphone cases, this shop offers a huge number, hung from walls and displays. Once at the counter, our attention is drawn to the repairer's tiny workbench. Located in an alcove, a high table with a non-slip green mat is home to various tools, charging connectors, as well as a jar of Nutella and a bottle of Sriracha spicy sauce. To one side, boxes are filled with tools and spare parts.

CASIO
TE-2200

iPhone 6/6s b/w
iPhone 6 Plus b/w

iPhone 6/6s
Zubehör
für
Samsung
Phone 5/5s

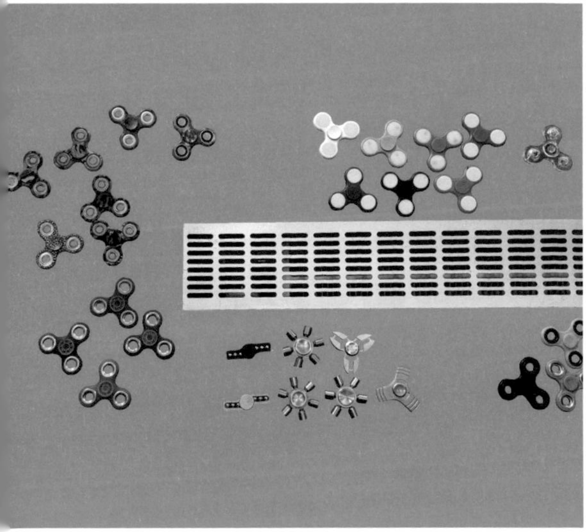

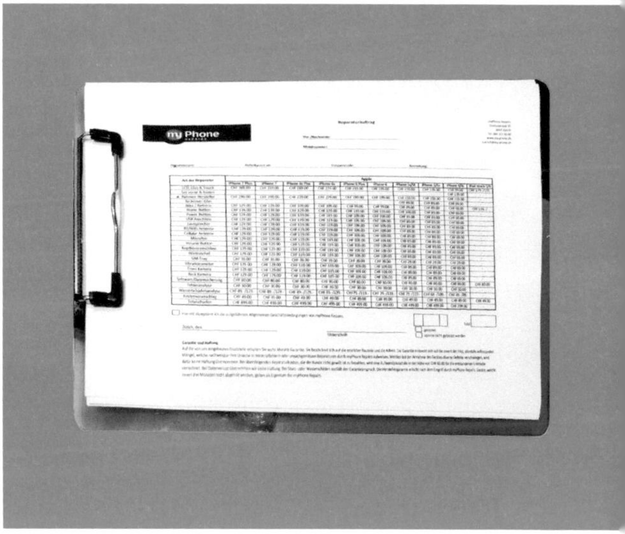
my Phone

A., (ZURICH): I started working here two years ago. I'm originally from Thailand, but have lived in Zurich for 25 years. I work here alone. The owners are two guys, one from Turkey and the other from Spain, but they rarely come here. They take care of all the administration, send emails to customers, and buy spare parts for repairs and accessories.

I worked at Interdiscount as an electrician. I knew someone who was looking for an employee in the workshop, and that's how I started to become a phone repairer. I was in the electronics department at the time, and one day, one of my friends asked me if I could fix his phone. I tried and it worked, so I sent my application to this store and started working here. I learned everything here from scratch. I first started at the workshop in Tivoli, where professional technicians showed me all the processes, for six months or so. For the first few months, my boss taught me how to fix things. Each day, I learned all the techniques and methods to work on the different models.

I find it difficult to be both the repair technician and salesperson. It's difficult to deal with customers, especially those who are frustrated. I have a lot of people who are upset here. In Switzerland, people are really stressed, you know, they never calm down. They wake up, work, sleep and work. Monday, Tuesday and Saturday are very intense days. Also, as the shop is rather small, I'm the only one they can complain to. In the same situation at Interdiscount, I would just say "sorry, this is not my department, you need to talk to the IT technician or sales manager." Here, you are responsible for everything. The work is easy when you have time, but it's not often the case. I like to have contact with customers. Sometimes I say, "okay, I can fix it," but I'm not sure. I can try for 30 francs, which is the cost of my working time. But if it doesn't work, they get angry.

When I need something or have a technical question, I call someone at the other store in Tivoli, the largest shopping centre in Switzerland. But I can do everything here except micro-soldering. In Tivoli, they have more repair equipment and two technicians. There, they also repair motherboards. When a phone has a problem at that level, I send it there.

PEOPLE COMPLAIN ALOT ABOUT THE TIME IT TAKES TO REPAIR THEIR DEVICES
Phone
NORMALLY IT TAKES ABOUT 30-40 MINUTES... BUT SOMETIMES IT CAN TAKE LONGER, OR EVEN FAIL!
HD

One day, a customer told me that he broke his phone by holding it into his mouth with his teeth while cycling! He had no pocket in his trousers.

JUSt like IN tHE GOOd Old dAYS

#10

"JUST LIKE IN THE GOOD OLD DAYS"

In an industrial building located in Kreis 4 in Zurich, blue plastic tubes placed on the protective grille of the ground floor windows form the letters 'Fab Lab', showing us the workshop's location. Inside, we are welcomed into a huge room. In the centre, mismatched chairs sit at tables, and shelves are filled with meticulously organized tools, machines and storage boxes. As we are contemplating the space, one group member begins to dismantle a concerned visitor's vacuum cleaner, examining it with various tools, before suggesting ways to restore it.

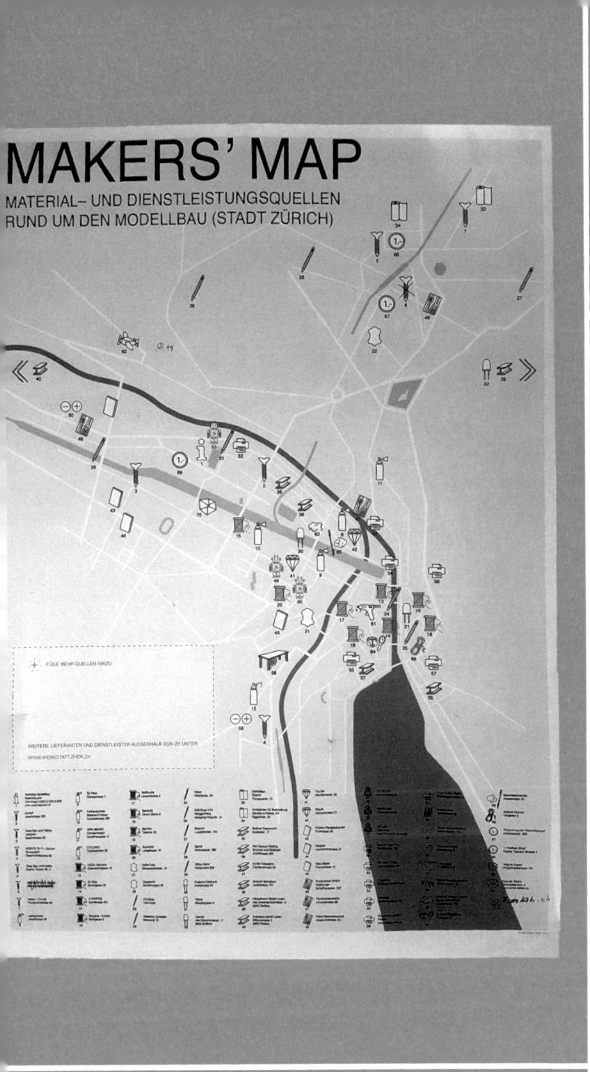
MAKERS' MAP
MATERIAL– UND DIENSTLEISTUNGSQUELLEN
RUND UM DEN MODELLBAU (STADT ZÜRICH)

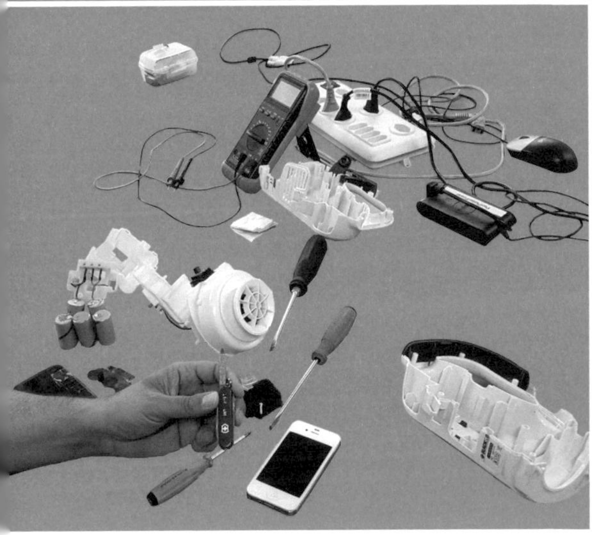

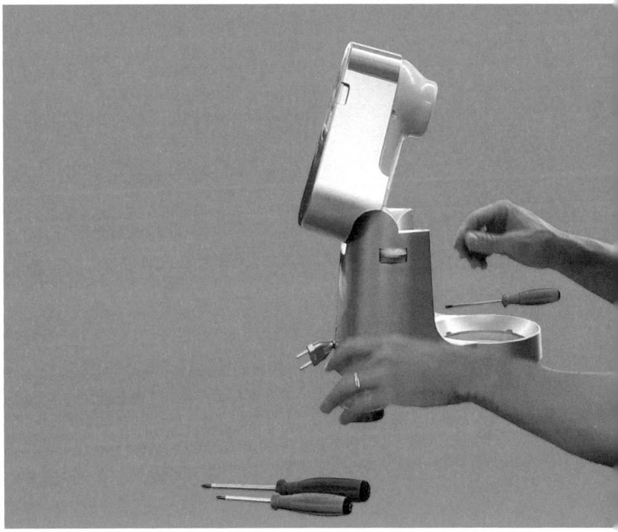

M., (ZURICH): I come from Eastern Switzerland, from St. Gallen. I moved to Zurich to study at the ETHZ for 4 years, that was 20 years ago. I previously did a more manual apprenticeship, I was also at CERN, and worked in the physics and mathematics department of the ETHZ. Currently, I work for a start-up developing processes for building robots.

The Fab lab is eight years old, and I am the person who organizes the repair cafés here. One day, I saw a television documentary about a repair café in the Netherlands. I figured we needed that here too! I asked the Fab lab people, we voted and everyone agreed that we should do it. We want to save the planet, it's a place where people can work together and repair together. Of course, you can also fix things at home, but it's more complicated without machines, and when you're alone. Here, you have access to tools and machines. And it works! We decided to remain discreet because if there are too many people, it becomes difficult to manage. That's why we don't do advertising to promote our events. We have different types of visitors, from the poorest to those people who really love their devices. For example, the other day, a man came in with a radio that he had bought with his first pay cheque, many years ago. We opened it up, cleaned it, and by adjusting the small capacitors, buttons and antenna, we fixed it, just like in the good old days.

The main goal of this kind of event is to help people get rid of their fear of electrical appliances. Most of the time, people say: "I can't do this." I just give them the tool and tell them, "Okay, now you have to open it," and they do it in front of me. The repair café is a moment of collective repair, all together. The problem with repairing phones is that they are glued and we don't really have the parts in stock. People often bring smartphones here, but we recommend they go to iPhone Doktor, a nearby store. The hardware part is easy to repair, we know how to do it, but the software is more difficult. With old iPhones, for example, there are many applications you can no longer use.

6
ATELIERHAUS
ZIMMERLISTRASSE
6
iFixit
RepairCafe
im
LAB LAB
ZURICH

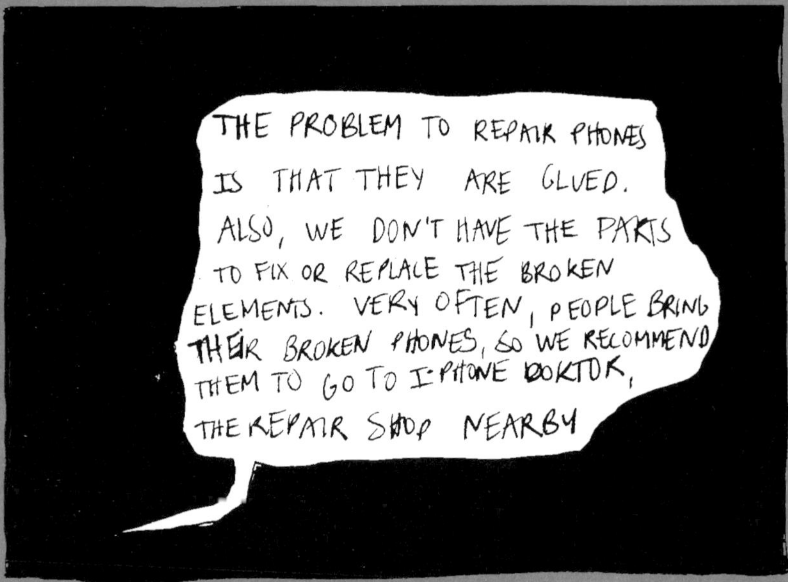
THE PROBLEM TO REPAIR PHONES IS THAT THEY ARE GLUED. ALSO, WE DON'T HAVE THE PARTS TO FIX OR REPLACE THE BROKEN ELEMENTS. VERY OFTEN, PEOPLE BRING THEIR BROKEN PHONES, SO WE RECOMMEND THEM TO GO TO I-PHONE DOKTOR, THE REPAIR SHOP NEARBY

LAB LAB
IPHONE DOKTOR

Repair kit + scissors + Swiss army knife

THE FIRST THING TO DO IS TO GET RID OF THE FEAR PEOPLE HAVE TOWARDS ELECTRONIC DEVICES. MOST OF THE TIME PEOPLE SAY "I CANNOT DO IT." I JUST GIVE THEM THE TOOLS AND TELL THEM "OK, NOW YOU HAVE TO OPEN IT" AND THEY DO IT!
YOU HAVE TO OPEN IT
← MINI-VACUM CLEANER

The guy thought that the condenser should get replaced.
M. then takes the opportunity to show us
how to micro-sold. The v then transforms
into a microsoldering workshop for the participants.

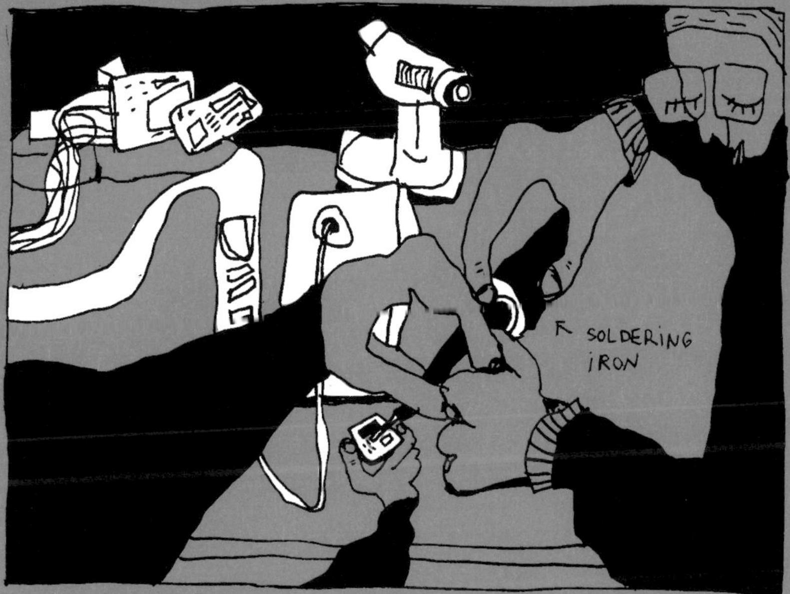

come
fix your
broken
items

#11

"COME FIX YOUR BROKEN ITEMS"

On the fourth floor of an industrial building, the O. association's workshop is one of the most densely-packed repair sites we encountered in our research. There is little empty space around us, shelves are filled with materials, with tools lined up on the walls, and machines everywhere. The central table is crammed with computers, papers, and other digital devices. There are only few gaps where you can slip from one corner of the room to another. The atmosphere is friendly, marked by care and effort, with everyone bent over electronic components.

MICROVERSE

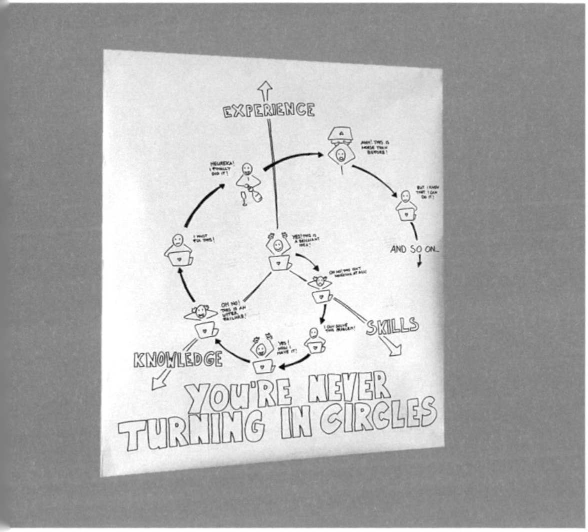
EXPERIENCE
AND SO ON...
SKILLS
KNOWLEDGE
YOU'RE NEVER
TURNING IN CIRCLES

RULES of PLAY
1. Enjoy and respect the tools and materials.
2. Clean your area and reset before you go.
3. Clean +1. Do one (more) thing to leave the space better than when you found it.

R., (RENENS): I'm a student at the Swiss Institute of Technologies (EPFL), studying electronics, but I only did some practical projects quite late in the curriculum, which is frustrating. My group of friends wanted to do something more concrete, more applied. We wanted to build a robot that we could send in a balloon to Antarctica—much like the stereotype of engineers looking for solutions to problems that don't even exist! It was the challenge that really interested us, specifically the way we could get there. We saw the value of such a project, and that's why we founded this association. We take courses at EPFL, and at the same time, we carry out our personal DIY and start-up projects in this workshop. Our members are often engineers, specialising in electronics, mechanics or microtechnology. But there are also people who study physics, math, chemistry, computer science or life sciences.

What matters here is the exchange between disciplines, the exchange of knowledge. It is the idea of creating a community where those who know more can show other people how to make things. We have about 50 members, but in 2014, at the very beginning, we were just five friends. Over the past three years, we have developed quite a few prototypes. For example, we started collaborating with glaciologists from Grenoble, in France, who found a use for our robot.

Most of our members come from EPFL, because of our network of friends, but here we are trying to create links with local associations and neighbourhood centres. We try to develop expertise around hardware and technologies. Part of the answer to this is in the tools, we want to establish tools and a community to promote mutual learning.

We're a bit like a fab lab, but we are not anti-start-up, unlike other Fab labs. We are here to help people with their technical problems.

We also organize repair cafés from time to time, because it's close to what we do. Basically, another local association related to sustainable development on campus invited us and computer experts (for software tinkering). At the repair café, we repaired headsets that people brought us, devices that no longer work because of poor cable contacts. In my experience, smartphones are not often repaired at these events. It's difficult. We would have to order the parts before each repair café, and that would be complicated to organize. But sometimes we have to look into that. For example, during the second repair café we organized in Lausanne, there was this girl who had bought smartphone parts on the internet. She came directly to us with the WiFi module to change, as she didn't want to do it alone. It took us two hours. An hour to dismantle the device, then another hour to reassemble everything. We had 35 compartments to separate the spare parts! That's what fixing technology is all about.

Repair café
T'AS DES OBJETS CASSÉS!
VIENS LES RÉPARER
29 nov17
COUTURE ET ELECTRONIQUE
CE 106
DE 18H15 À 22H
Cinderella

AND SUDDENLY, SOMEONE YELLED "AH I LOST A SCREW!" IT IS STRESSFUL WE WERE THE LAST TO LEAVE. THE OTHERS WERE LEAVING BUT WE COULDN'T WE HAD A DISMANTLED PHONE TO REASSEMBLE!
AHH
NOOOOOO
I LOST A SCREW
ARRGH

WE DON'T FIX SMARTPHONES, THEY'RE TOO TRICKY TO OPEN. THERE ARE OFTEN MORE THAN 40 STEPS TO FOLLOW AND THEN... WE'D HAVE TO ORDER SPARE PARTS BEFORE EACH REPAIR CAFE, IT'S TOO COMPLICATED.
THAT SAID, AT THE SECOND REPAIR CAFE WE ORGANISED, A GIRL BOUGHT A COMPONENT THAT SHE WANTED TO REPLACE. IT TOOK US TWO HOURS. AN HOUR TO OPEN THE PHONE AND DISMANTLE EVERYTHING
AND ANOTHER HOUR TO PUT IT ALL BACK. WE HAD 35 BOXES TO KEEP THE PARTS SEPARATE!

FAB
I CAN'T FIX A SMARTPHONE. I KNOW ELECTRONIC COMPONENTS THOUGH, IT'S A QUESTION OF INTUITION. MY APPROACH IS GENERAL, I DON'T KNOW THE DETAILS... THERE ARE WEBSITES SUCH AS IFIXIT, I GUESS YOU KNOW THAT. YOU LEARN THROUGH PROJECTS. YOU UNDERSTAND THAT YOU CAN'T FORCE THINGS.
YOU HAVE TO BE PATIENT
A RACK

We learn best by making mistakes.

Alone but in a network

#12
"ALONE BUT IN A NETWORK"

Squeezed between the television and computer, tablet and mobile phone departments of Geneva's B. shopping centre, under a sign indicating "express repair," Phone Solutions' repairers occupy a small kiosk, with just enough space to move from the repair space to the cash register. In this area of high footfall, facing the escalators, it is impossible to miss its presence. The workspace is reduced to its simplest expression, as if it had been optimized to best slot into an already densely occupied space.

Phone
Solutions
RÉPARATION
EXPRESS
SMARTPHONES
TABLETTES

MILOS RAONIC CAN 7 6
MARCO CECCHINATO ITA 6 3
Phone
Solutions
RÉPARATION
EXPRESS
SMARTPHONES
TABLETTES
logitech

F., (GENEVA): I am a manual person, I have always tried to understand things, to improve and optimize everything. I have never been afraid of opening an object to use it better. I trained as a welder. I did a BEP in Annecy, in France. My passion is automotive mechanics, but I am self-taught with smartphones. I've always liked putting my fingers inside machines. It is important to understand what you are doing, but also to find out what is happening, and why. It's important to feel it with your fingers. I taught myself when my camera broke, trying to make it work again.

When I started working here, I had 2 months of training. I worked with another technician, and then alone, all by myself. The company is managed by two partners, my bosses, who have opened local kiosks in shopping malls across this region. We are an independent company, we are not part of Manor. Our store is not approved by Apple, but that will happen soon. There's a whole process for that, you have to do some training, follow their rules. In total we have about 20 employees, including the people in charge of the office, and inventory management. We hardly ever meet with the whole team, but we are constantly connected with WhatsApp. We have one group per kiosk. The work is sometimes difficult, you have to manage customer expectations, sometimes they argue with you. It's still technical but there's this whole customer part that I didn't need to do before. It is often stressful to be exposed to customers, it is difficult to manage information and repair at the same time. There are two of us at the Balexert shopping centre. I am a nomad, sometimes I am in Manor, or Balexert or Chavannes.

At this point, I don't know if I want to remain a repairer. It's an interesting job, but it's not really considered as a profession. In terms of salary, we earn the minimum, we earn as much as someone who is a cashier while we have real know-how. We need to be able to identify if the device will lose its data, if it will work again after the repair, if it will be restored in time… It's quite a skill. Sometimes, we have to do welding, but we don't do that here, we have a dedicated centre in Lausanne. Depending on our customers' problems, there can be a long waiting period, sometimes a month. This is mainly because we have to order parts, and depending on the brand or model of phone, it can take a long time. We have suppliers all over the world, in Europe, Asia, the Middle East (Dubai), but even if we meet suppliers from our network at trade shows on a regular basis, it takes time to import their parts.

AS SOON AS A NEW MODEL COMES OUT, WE HAVE TO UPDATE OUR KNOWLEDGE. WE BOUGHT AN IPHONE 7, OUR TECHNICIANS BROKE IT. IT COST US 1000.- BUT IT WAS AN OPPORTUNITY TO LEARN
EVERY PHONE IS DIFFERENT. FOR INSTANCE, THE IPHONE 7 DOESN'T HAVE A HOME BUTTON. ONLY APPLE CAN REPLACE THEM. WE HAVE TO ADAPT MANUFACTURERS' STRATEGIES, THEY DO EVERYTING THEY CAN TO KEEP US FROM FIXING THEIR PRODUCTS

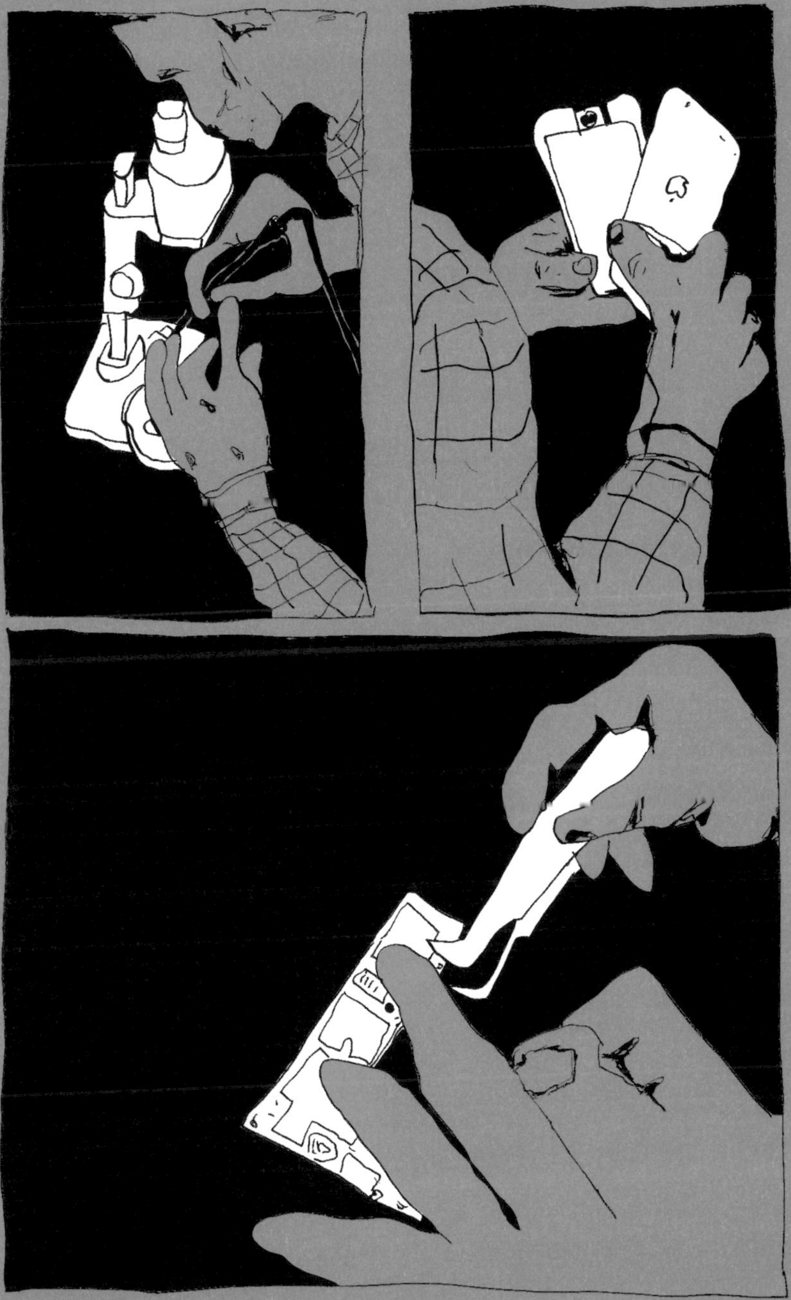

F. shows me the WhatsApp group where sales people and technicians discusss various problems, techniques, spare components, and stock schedules.

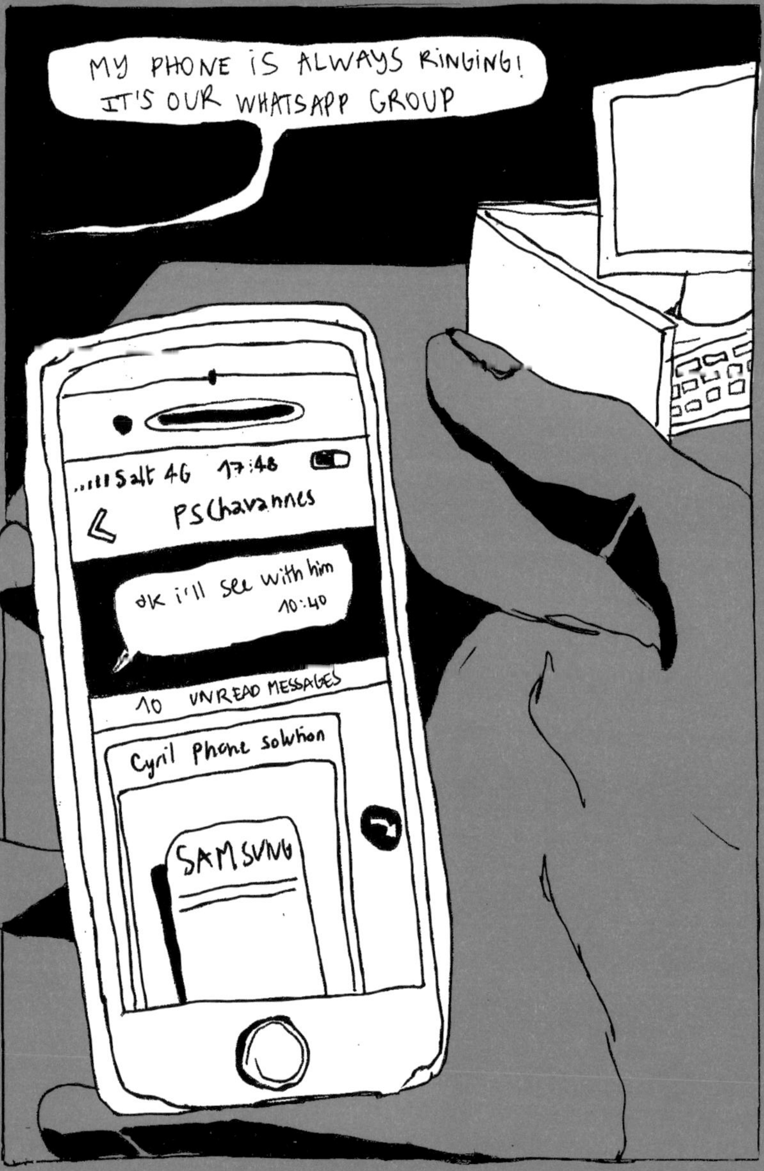

WE ONLY EXPLAIN THIS TO CUSTOMERS WHO USE ANOTHER SERVICE OTHERWISE WE'D SPEND OUR TIME UPDATING PEOPLE'S APPLICATIONS!
Computing
Phone solutions
Phone solutions
EXPRESS SMARTPHONES

hama
hama
hama
WE OFTEN GIVE CUSTOMERS TIPS, HOW TO BETTER USE YOUR SMARTPHONE, THE BATTERY, APPS, BLUETOOTH ...
ONE Solutions

we arE
liKE A phar-
mACY

#13

"WE ARE LIKE A PHARMACY"

Between a kebab shop and a strip club in Geneva's Paquis district, the ground floor of the building, comprising two different shopfronts, has just reorganized: the hairdresser and the repair shop have swapped places. There had not been much of a clear separation before the switch, with customers' phones being serviced before a haircut in the salon. The shop is as lively as the neighbouring salon, and customers keep coming in, asking for information, or just saying hello and chatting with the team of repairers. During our research, the shop did not just change premises, but its logo and overall front design evolved to become more refined, reflecting its growth. A., the Iraqi manager, receives us, indicating that he set up the business on his own.

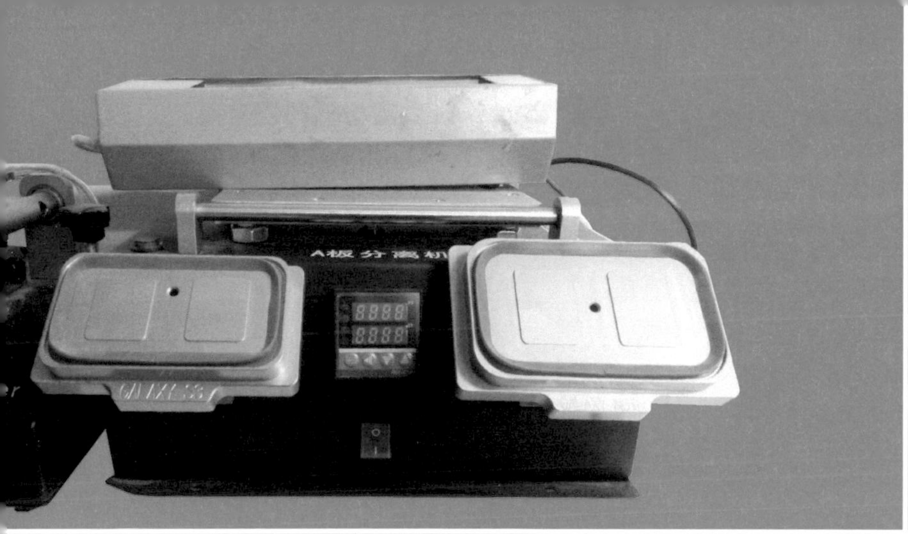
A板分离机

分离贴合机

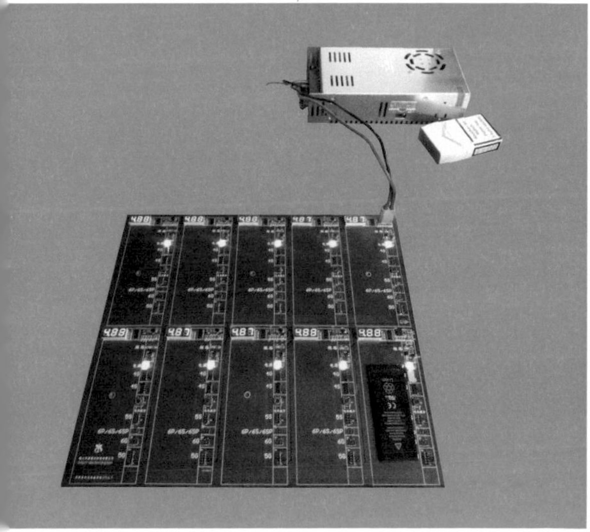

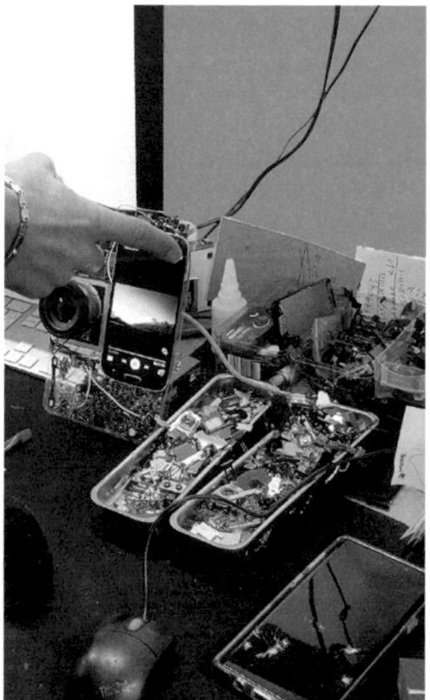

A., (GENEVA): When I was a kid, I liked electronics, opening things and repairing them, I was drawn to technology. Before Classphone, I worked in the food business. I had a restaurant in the neighbourhood but as everyone could see I was attracted to technology, I was advised to follow that. The shop is seven years old. The first year, I had four stores, three in Geneva and one in Bern. I did everything myself, I would collect the parts on my scooter and take them from one store to another. In Bern, they sent me the parts by post, I ran all over the place and never slept. Then I hired people, and the business grew. Over the past two years, it's been getting more difficult. People want to invest less. Now I am the only person in the shop, I ask people to cover for me when I'm travelling, but otherwise I'm alone. I had four employees, before, just in this shop. But it was too stressful, I couldn't sleep at night. Now I'm alone and it's going very well, I am calm, I don't have salaries to pay each month. Anyway, in this area, if the boss is not in the shop, it doesn't work. So you have to be there all the time. All the time. Otherwise, things don't go as you want them to.

Success is a question of strategy. You need to have what people want. It's like a pharmacy, if you don't have aspirin, no one will come. So I immediately started buying spare parts. Today my parts come from Germany, Poland, Iraq, and some from Dubai, where they are cheaper. In fact, in this business, you have to find where the parts are cheapest. I also go to London, where I have my suppliers, four times a month. In London, everyone is doing this, depending on the neighbourhood, it's Indians, it's the English. I also work with French people in London. On the other hand, I don't buy anything in China, I can't trust the quality, I don't have suppliers there. I worked with someone there in the beginning, but the quality was not good. And then there are the costs, it's complicated. In China I just buy accessories.

Class
Coiffure

BALENCIAGA
PARIS

Hi CHIEF
SALAM SALAM
HEY COUSIN, DO YOU HAVE AN S8?
HOW MUCH?
HE IS NEW!
COULD BE 200.-
GIVE ME A DISCOUNT
OK OK
HEY BROTHER
WALLA, THANKS

GOOD MORNING, I HAVE A PROBLEM
I HAVE THIS OLD COMPUTER

THE SCREEN HAS TURNED
INTO AN AQUARIUM.
CAN I BRING IT OVER?

BALENCIAGA
PARIS
SCREEN
AQUARIUM
Dell

HELPING PEOPLE

#14
“HELPING PEOPLE”

Located on the fifth floor of a building on Bahnhofstrasse—Zurich’s well-known shopping street, which connects the lake with the main station—the shop looks more like a clinic or the office of a multinational company than a repair store. Stuck between accounting firms, investment funds, and private bank offices, HandyDoktor seems to have adopted the architectural codes of its neighbours. Inside, a reception area with leather chairs opens onto a counter and a workshop hidden behind a large, opaque screen. A woman busy fixing a Samsung Galaxy receives us.

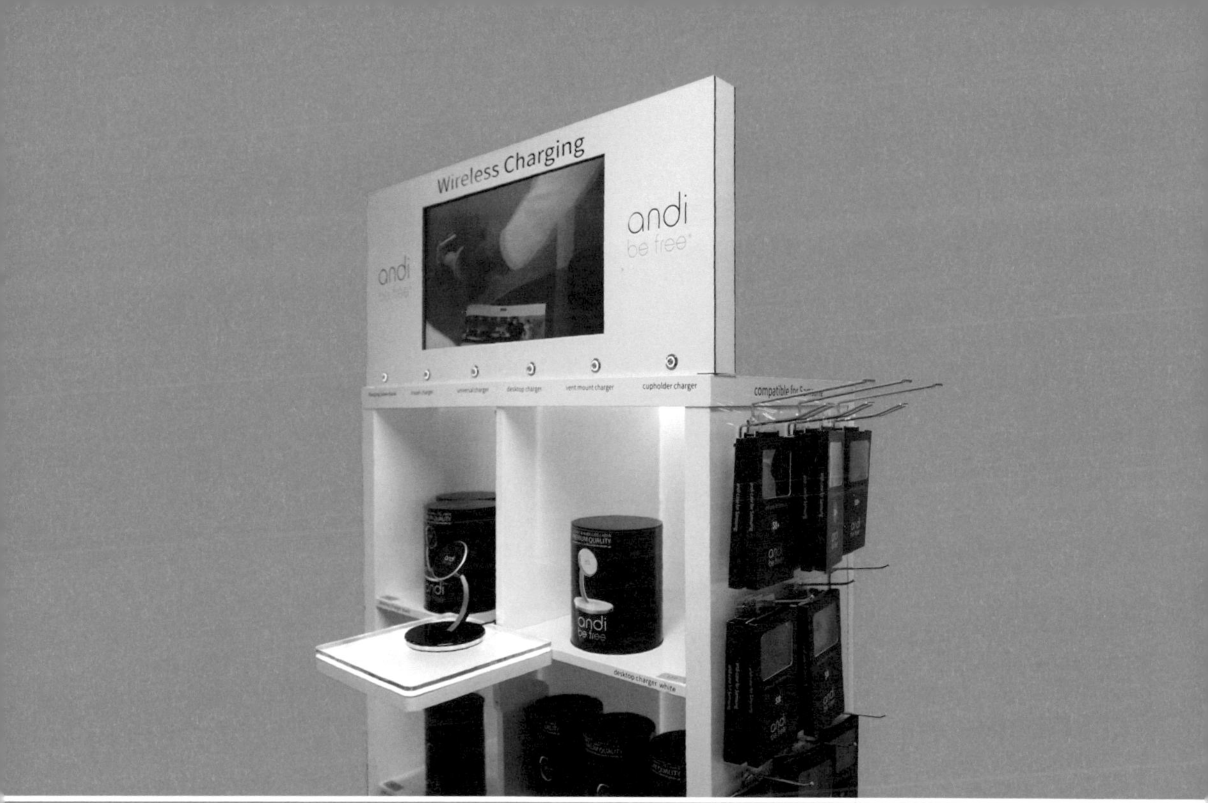
Wireless Charging
andi
be free
universal charger
desktop charger
vent mount charger
cupholder charger
desktop charger white

M., (ZURICH): I come from Russia, I married a Swiss man and we moved here nineteen years ago. It was my first job in this country. A friend of was working here and invited me to join the team, that was four years ago. Work is sometimes boring, women like me need a more lively life. And I have the impression that men feel better here than women. The longer you stay, the more boring it becomes.

When I came to work with my colleague, she taught me everything. After 6 months, you are pretty good, you know the important things. It was difficult at first, because I was afraid of breaking parts. It was hard and I had to explain this risk to people. The worst part was when I didn't know the answer, because I had just started. At first, I felt that I was not efficient enough, but it was difficult to know everything, and I could not answer all the requests. Another thing I soon noticed was that people accept when men don't have an answer, but not when they talk to a woman.

I do different things here, it's a balance between the technical side and a desk job, but I like that. Sometimes I get tired of doing nothing but repairs, it can be tedious to always have my nose stuck in these devices. And I like working with customers, doing different things. In addition to the technical side, I am responsible for the organization and administration of this shop.

My two bosses are Swiss, they opened this store five years ago, when there were not many places like this. Three of us work here. I help customers, and a colleague does the more complicated repairs. But I want to change jobs, that's why I am studying psychology, I take courses online, on a Russian platform. This work made me realize I like helping people. Here, you help a lot, all day long! That's why I'm interested in psychology.

Handy Spital.ch
SOMEONE ONCE TOLD ME "I FEEL THAT SOMEONE PUT SOMETHING IN MY PHONE TO FOLLOW ME". BUT THERE WAS NOTHING WE COULD DO. HERE WE DON'T DEAL WITH SOFTWARE ISSUES... IT IS TOO COMPLICATED AND EXPENSIVE!
Handy Spital.ch

THE IMPORTANT THING ABOUT THIS JOB IS TO BE CAREFUL BECAUSE YOU CAN MAKE PEOPLE REALLY UNHAPPY. THEY LOVE THEIR PHONE! FOR EXAMPLE ONE TIME A PERSON LOST ALL THE PICTURES OF HER BABY!
cafee

WE ARE TINKERERS

#15
"WE'RE TINKERERS"

Occupying the first floor of a former printing factory that is being converted into a startup incubator, the space is different from the building's other sections. Unlike the neighbouring offices, the place's purpose is immediately obvious, with the sheer density of material in the room showing that we are in a hackerspace. A phenomenal amount of machines, tools, and storage cabinets seem to be available for all types of projects.

The furniture is varied, and a range of signs give advice to new or occasional visitors. Next to the sofa there is an arcade terminal, leaning against shelves stacked with manuals and books of all kinds. A novel by author Bruce Sterling catches my attention. A sensor on the door is connected to internet, announcing on Twitter if the space is open, and signalling to potential users that they can come and make use of the space. The presence of stickers on most of the appliances in the room gives the impression these objects are begging for us to repair them. We're definitely in a hackerspace.

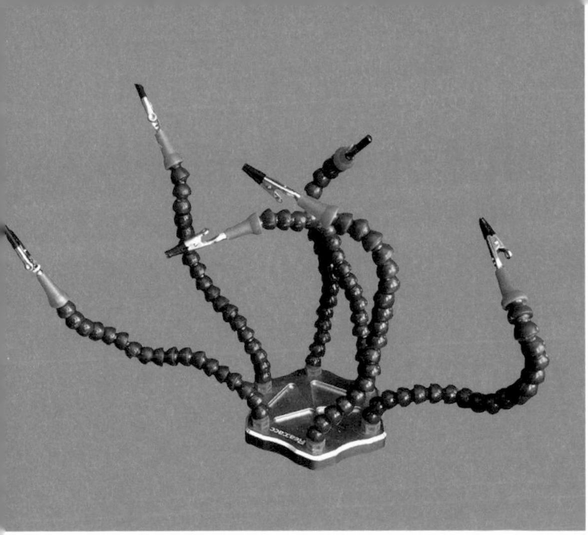

Nano
Arduino
Uno
LM 566
LM 393

TRON

ELECTRO TRASH

S., (RENENS): I studied computer science. Currently I work as a freelancer, I'm on a project with three friends who are starting their own business. I discovered this hackerspace seven years ago. I had heard about it on a podcast and I was curious. Then I came back a few times to tinker with things and learn. After a while I wanted to get more involved, and that's why I joined the committee last year. Today, I am president of the association. For a long time I had no concrete projects of my own, I came to learn more by watching others and helping.

We are an association with sixty active members and another twenty people who come less often. We reach many people but most of our members are interested in computers or students with a project, unlike the fab lab in the building below, which reaches different people, such as designers, architects or artists. If an object is broken, at the fab lab downstairs they look on the internet for information before trying to repair it, they call the manufacturers, etc. We are less cautious. We want to solve the problem as quickly and efficiently as possible, without trying too hard to understand the system as a whole, or spending days reading documentation. The problem is that, in doing so, there is sometimes a risk of breaking other parts of the object!

Speaking of broken devices, we have members who often repair their own phones. For example, I fixed my iPod touch 4 when its mini jack was no longer working. I had ordered the replacement parts and found information on the internet. There were other members watching and helping me. Everyone walks around the workshop and looks at at what people are doing, the solutions they adopt, wondering "why do you do it like that and not like that?" It's important to get feedback when you're tinkering. It's an exploratory process.

The fab lab downstairs has facilities for manufacturing, 3D printing, laser cutting. They focus more on technical inventions and industrial methods, working to help their members find solutions to particular problems. Here, it's more about helping make a circuit to control LEDs, using everyone's knowledge to tinker with the resources at hand. We have the same basic philosophy as the hacker fab lab and we also share equipment. Besides it's mostly instruments and servers that we found in other places, from people who wanted to trash them.

WE HAVE AN OLD ELECTRONIC WORKSHOP WITH OSCILLOSCOPES. MOST OF THE INSTRUMENTS AND OUR SERVERS ARE MACHINES WE FOUND. FOR INSTANCE, WE REUSED THE FORMER SERVERS FROM THE UNIVERSITY
CYBER FRIDG
ONE OF OUR MEMBERS WORKS AT THE RECYCLING CENTER. WE KNOW WHAT PEOPLE THROW OUT. LOOKING THERE, WE OFTEN FIND THINGS, TOOLS AND STUFF
ELECTRO TRASH

YOU KNOW, OUR MEMBERS OFTEN FIX THEIR OWN PHONES HERE!
TRO

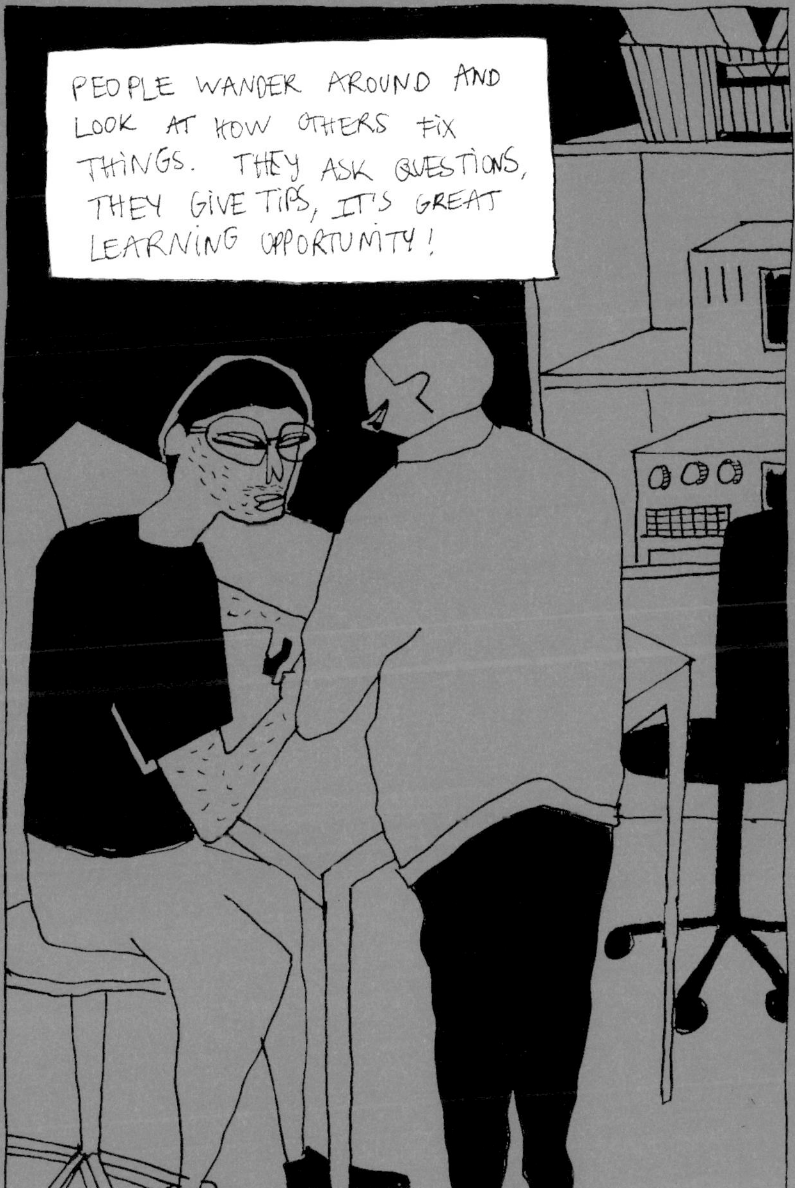
PEOPLE WANDER AROUND AND LOOK AT HOW OTHERS FIX THINGS. THEY ASK QUESTIONS, THEY GIVE TIPS, IT'S GREAT LEARNING OPPORTUNITY!

With many electronics devices, we wonder how to reuse them.

YOU HAVE PEOPLE'S BABIES IN YOUR HAND

#16
"YOU HAVE PEOPLE'S BABIES IN YOUR HANDS"

One Friday morning, about 30 minutes after breaking my iPhone—which I had carelessly left on the edge of a photocopier—I go to the first store to open so early in the day. P.H., on the Plaine de Plainpalais seems to meet my needs, as indicated by the sign on the sidewalk promising "iPhone and Samsung repair on site." I take a seat on the chairs in front of the counter, alongside the other early-morning customers, looking at their shoes, if not their smartphones, and wait my turn.

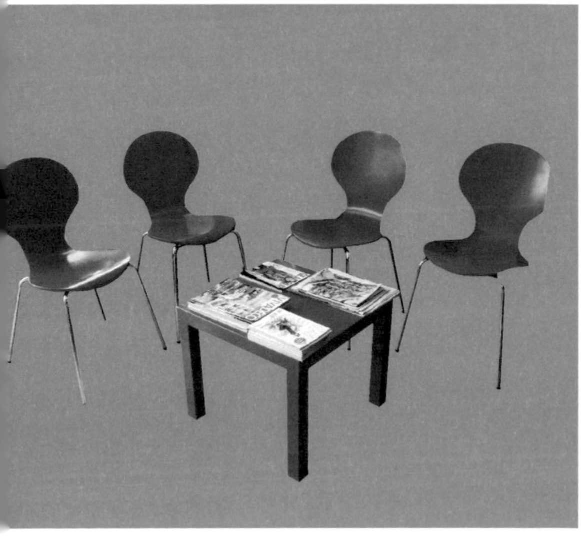

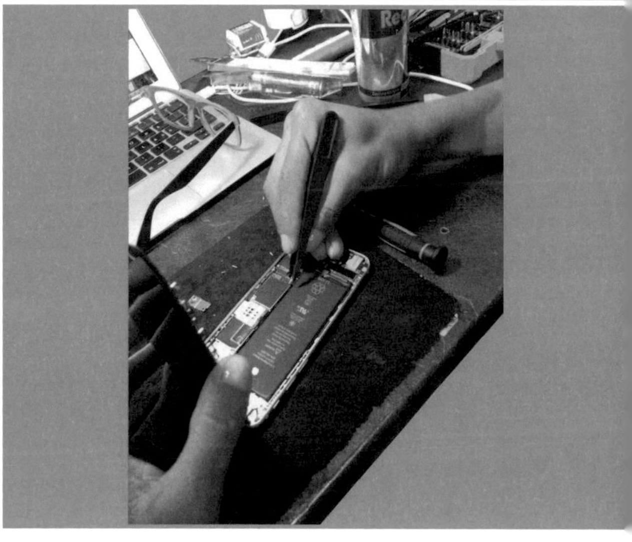

A., (GENEVA): I'm the manager, but I work here two days a week. I am Caribbean, I come from Lyon. In the store we have a repairer who trained as a computer technician, another in telephony, and extra people when the main repairer is not working. It's been three years since we opened the shop. That was my idea, because I've been in the telephone business for eight years. I have always worked part-time, for example, at Cash Converters, where I am still working today.

Why open a repair shop? We have a lot more freedom of thought and action. In supermarkets, there are fixed rules, but here we have the opportunity to buy new products, offer new services, etc., and this is related to our customers' environmental concerns. At first, I suggested opening this store to my boss at Cash Converters, but I finally did it myself. We work in partnership with them, we repair their products and they send us their customers who bought second-hand phones. I send them people wanting to buy products. We specialize on iPhone models, but sometimes our customers bring us smartwatches, iPods, computers, network peripherals, and so on. We plan on setting up a computer repair service to meet these customers' needs.

I learned on the job, even though I have a management background. In this field, that's quite common. We work with three repairers, all of whom are part-time employees, with an hourly contract. C., who is here today, has a diploma in management (BTS) from France, and also works as a salesman at Cash. E. is the main repairer. He is versatile and can make both sales and repairs. He works full-time. Then there is J. who comes when C. cannot replace E. I can't give him a contract because he has a different status, he is Spanish.

C., (GENEVA): I started repairing with my father when I was 5 years old! He was repairing computers and I started doing it with him. At one time I tore everything down, all the devices. I live in Aix Les Bains, in France, and I travel to repair phones in Switzerland. Basically, I found an ad on Anibis. The manager initially took me on as a consultant. And since she was happy with my work, she hired me, and now I even work at Cash. I'm also a musician, so I manage the music department there. I fix everything. You have to be a bit of a McGyver to do this job. You have people's babies in your hands—I like the feeling of helping people. Then there's the discovery side. To be able to repair something, you need to understand all the parts. Knowing what tool to use and when, I like that.

SHOP OWNER
CUSTOMER
REPAIRER
PHONE

1.
94%
2.
3.
4.
5.
6.
7.
SELFIE

EVEN APPLE TINKER WITH THEIR PHONES YOU KNOW, I HAVE AN iPHONE SE WHEN I OPPENED IT, I SAW THEY ADDED PARTS FROM AN iPHONE 5, A PROCESSOR AND A CAMERA FROM THE 6S. IT'S SUPER FAST.

I SAW THEY ADDED PARTS FROM AN IPHONE 5,
A PROCESSOR AND A CAMERA FROM THE 6S.
iphone 6
magnetic screw mat
IT'S SUPER FAST

DO YOU KNOW WHAT JAILBREAKING IS? IT'S ILLEGAL 'COS IT LEADS TO DATA CORRUPTION
MOBILE
TABLETTE
IT'S A PROGRAM THAT EXPLOITS A FLAW IN THE SYSTEM.
IT'S LIKE PHONE CANCER.
YOU NEED A PARTICULAR PIECE OF SOFTWARE AND A COMPATIBLE MACHINE FOR THAT, THOUGH

tHe
sOurce

#17
THE SOURCE
华强北 [HUA QIANG BEI], 深圳 [SHENZHEN]

From a distance, Hua Qiang Bei looks like a series of large, fairly generic buildings.

From the street, the small shops in this large commercial district seem to overflow their premises.

Inside, the further you press on, the more you discover the size and scale of the whole. If the shops on the ground and lower levels resemble the classic stalls of electronics stores, a visit to the upper floors reveals a multitude of small signs, sometimes reduced to their simplest expression, stands selling highly specialized electronic components: microprocessors, motherboards, cables, sensors, microcontrollers, screens. In the corridors, which narrow as you progress through the structure's labyrinthine architecture, people seem out of sync: some mill around, while others seem to proceed straight to their goal, rushing to a particular stall or expert. Employees spend lunch breaks on their smartphones, snooze, pet their dogs, or patiently show price lists to potential customers who can't read the language. Westerners are amazed by the profusion of bric-a-brac. Some people do business, while others wait for clients.

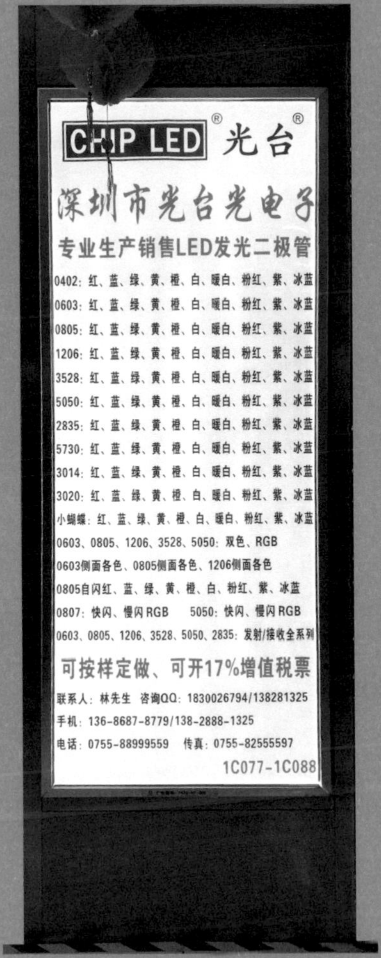
CHIP LED® 光台®
深圳市光台光电子
专业生产销售LED发光二极管
0402：红、蓝、绿、黄、橙、白、暖白、粉红、紫、冰蓝
0603：红、蓝、绿、黄、橙、白、暖白、粉红、紫、冰蓝
0805：红、蓝、绿、黄、橙、白、暖白、粉红、紫、冰蓝
1206：红、蓝、绿、黄、橙、白、暖白、粉红、紫、冰蓝
3528：红、蓝、绿、黄、橙、白、暖白、粉红、紫、冰蓝
5050：红、蓝、绿、黄、橙、白、暖白、粉红、紫、冰蓝
2835：红、蓝、绿、黄、橙、白、暖白、粉红、紫、冰蓝
5730：红、蓝、绿、黄、橙、白、暖白、粉红、紫、冰蓝
3014：红、蓝、绿、黄、橙、白、暖白、粉红、紫、冰蓝
3020：红、蓝、绿、黄、橙、白、暖白、粉红、紫、冰蓝
小蝴蝶：红、蓝、绿、黄、橙、白、暖白、粉红、紫、冰蓝
0603、0805、1206、3528、5050：双色、RGB
0603侧面各色、0805侧面各色、1206侧面各色
0805自闪红、蓝、绿、黄、橙、白、粉红、紫、冰蓝
0807：快闪、慢闪RGB 5050：快闪、慢闪RGB
0603、0805、1206、3528、5050、2835：发射/接收全系列
可按样定做、可开17%增值税票
联系人：林先生 咨询QQ：1830026794/138281325
手机：136-8687-8779/138-2888-1325
电话：0755-88999559 传真：0755-82555597
1C077-1C088

天信

Hua Qiang Bei is the world's largest electronic market. Sometimes referred to as the "Silicon Valley of Hardware," the shopping centre is a hub of the electronics industry. In the open space of a fifteen-storey building, merchants and technicians refurbish and sell used electronics. Each stand specializes in a particular range of components, a type of machine, or a specific brand. On site, thousands of craftsmen and merchants dismantle, clean, sort, test, assemble, display, exchange, pack, and send electronic equipment elsewhere on the planet. It is also in this area that used and damaged smartphones are collected, repaired, or dismantled, for use in the renovation of electronic components.

B., (GENEVA, LYON): Every three months we went to Paris, to shop for spare parts and accessories. We realized that these components were our biggest expense. So we thought, why not find the best supplier and become our own reseller? We realized that those suppliers were in China. In 2008, I went to Hong Kong. At the time there was no internet, and so it was difficult to find suppliers' addresses. We just knew it was in Shenzhen. We walked ten miles to find out, we crossed the border. At the time, there were six or seven million inhabitants and no one spoke English. In Hua Qiang Bei, the good thing is that I started at the same time as everyone else. There were 800 to 1000 stores. The more you went down into the buildings, the more expensive it was. I noticed it after four or five trips.... Because of the rents... the prices are more expensive down there, and when they see that the customers are serious, they make them go up. It's changed since then, with Alibaba. Over time, we became friends with our suppliers. To trust each other, you really have to be friends, and in China they don't trust anyone at first. Now it's not even work relations. When you go there, you eat with them.

DISCUSSION
WHAT CAN WE LEARN FROM THESE PORTRAITS? WHAT DOES OUR COMPARISON OF THESE REPAIR SPACES TELL US ABOUT REPAIRERS' LIVES? WHAT CAN WE LEARN ABOUT BROADER PATTERNS OF MAINTENANCE FROM A STUDY OF SMARTPHONE REPAIR IN SWITZERLAND?

In this concluding chapter, we seek to address these questions, drawing together the experiences depicted in the portraits, along with additional materials from our research.

LANDSCAPES OF MAINTENANCE AND REPAIR

Early in our research, we saw that the apparently clear-cut hackerspace/store distinction concealed a diversity of repair venues. While hackerspaces are usually set up as non-profit organizations (#10 #11 #15),[1] we also found entrepreneurs seeking to integrate commercial and non-commercial activities (#11). Equally, some for-profit stores had adopted a cooperative structure (#2) or a business model that struck a balance between generating revenue and serving the community (#8). As one of our interlocutors in Lausanne stressed, hackerspaces do not have a monopoly on wanting to "save the planet," with many stores stressing similar environmental objectives (#11). However, even if some of the businesses and hackerspaces we visited had broader goals, most repair shops still treat smartphone repair as a skilled commercial activity, based on their identification of user needs as a business opportunity. Our analysis of repairers' motives and employment histories supports this reading, as questions of sustainability and environmental issues were only referenced by informants from the professional classes, and those who had completed higher education.[2]

The shops we visited varied in size; from the supermarket repair-booth (#12) to the small neighbourhood store (#1 #3 #5 #9 #13 #14 #16), from the chain of shops operating across several Swiss cities (#6 #7 #8) to the sole proprietor (#4). Regardless of a given business's legal structure or scale, the set-up was normally the same, with one or two employees active on the shop

1 These reference numbers correspond to the previous portraits.
2 This is not to say our other subjects never talked about environmental issues, but such were not mentioned as one of their primary motivations.

floor,[3] covering reception, offering advice, handling payments, and making repairs.

Although different firms' motives and operational scope varied the repairer's objective remained the same: solving the problems encountered by their customers with their smartphone.[4] With problems and interventions often explained by analogy with human health and medicine,[5] store names often reference the medical and caring professions[6] (as seen in the introduction). Where Fab labs and hackerspaces organise "repair cafés," commercial repairers describe their interventions as a form of therapy (#3) or medication (#13). Devices are "sick," suffering from "Apple Loop Disease"[7] or "Touch disease,"[8] and repairers regularly use terms like "diagnosis," "infection," "viruses," "life expectancy," "resuscitate a smartphone," "bring it back to life," etc.

Beyond this desire to "cure" devices, the kinds of repair activities carried out in these spaces vary. As our interlocutors based in hackerspaces or Fab labs attested, smartphone interventions in non-commercial venues are extremely limited. These repairers' scope of action is limited by the complexity of repair tasks and a lack of spare parts. These factors can frustrate participants in collective repair sessions, who often turn to these spaces as substitutes for a paid service,[9] having heard the "Maker revolution" or "Fab lab phenomenon" framed as a means to combat object obsolescence. As in the portraits, the difficulty of repairing digital devices, as compared with mechanical technologies like vacuum

3 Diverging from the larger groups of informal repairers active in Bangladesh (Jackson et al. 2014) and Port-au-Prince (Payen et al., 2019), and the craftsmen involved in refurbishing smartphones in Huaqiangbei (Schulz, 2017).

4 Supplementing their income with the sale of phone cases, screen protectors, cables and adapters, alongside other electronic gadgets.

5 With some businesses even going so far as to incorporate such metaphors in their space and choice of décor. Although watchmakers, garage mechanics, and shoemakers work on objects that are also common, few repairers make such easy recourse to the lexical field of health. With the technique increasingly seen as an extension of the body—as anthropologists such as André Leroi-Gourhan (1964) have shown—wouldn't this medical metaphor reflect the importance of the technical object it concerns, and the affects it mobilizes? (cf. Nova, 2020)

6 Switzerland's iklinik repair chain features the electrocardiogram "pulse" as decoration on its walls, while labelling the different areas of the shop with terms such as "operation table."

7 https://www.vice.com/en_us/article/8xem8x/iphone-7-grayed-out-speaker-loop-disease-boot-loop-hangs-on-apple-logo

8 https://www.lifewire.com/iphone-touch-disease-4120914

9 Another hackerspace in Geneva, whose portrait we have not included in this book, indicated that they regularly received visits from people who thought they could have their smartphones repaired, following radio or television shows on Fab labs.

cleaners or toasters, see users quickly reassess the feasibility of re-appropriating the smartphone.[10]

For these reasons, shops remain the primary venue for smartphone repair. Most hardware interventions are made in response to customers' requests to change broken screens or mechanical parts (buttons, speakers, chassis), replace batteries or antennas, or dry out the device after it has been immersed in water. This allows us to draw an initial dividing line between different types of hardware interventions. The bitter words of one shop owner (#1) registered a difference between stores that "only know how to repair screens" and "real technicians." Where the former group are usually self-employed entrepreneurs who see simple repairs as an easy source of income, the "real technicians" are those repairers who can address complex or uncommon problems, or who possess advanced skills (#4 #5)[11] and techniques for diagnosing the causes of breakdown or failure. This distinction is not just a question of individual repairers' skills or specialisation, but derives from a way of valuing specific know-how, with certain technicians recognised as experts in particular practices (see Houston, 2019, p. 135).

A second divide separates repairers who focus exclusively on hardware from those involved in both hardware and software interventions. Within software interventions, there is a further distinction between operations that are easy to perform (e.g. OS updates and configuration) and more complex tasks, like data recovery, which may require computer programming skills. In addition, an emerging practice sees repairers intervene in the "behaviour" of algorithms, apps, and software. This might involve recommendation systems or voice assistants, such as teaching the Siri virtual assistant to recognise the user's voice, correcting pronunciation errors, or calling third-party services such as Uber.[12]

10 As the founder of a prominent Geneva hackspace told me on Twitter, responding to a link I posted on a public initiative on the "right to repair:" "With multilayer circuit boards with internal components, repairability is impossible. Having the right to repair is irrelevant if a) the device cannot be safely disassembled and b) the parts needed to repair it cost more than a new device."

11 Such as micro-soldering (soldering operation that requires using a microscope in order to connect wires or small-scale components to each other with an alloy made from tin and copper), or screen reconditioning.

12 A phenomenon close to what communication researcher Tarleton Gillespie calls "tacit negotiation," when describing how "we perform to adapt ourselves to algorithmic systems: we enunciate differently when speaking to machines, use hashtags to make updates more machine-readable, and describe our work in search engine-friendly terms." (Finn, 2017, p. 20)

FRICTIONS IN REPAIR PRACTICE

Given the varying objectives, organisational structures, and degrees of commercialisation of these different repair sites, our portraits capture several frictions key to these spaces' everyday lives.

The most visible tension is in the precariousness of most of these spaces. Apart from the Swiss German retail chain (#8), most of the repairers, owners, managers, and employees we spoke to report that repairing can be a tedious activity. Depending on a store's location, customer demand is sometimes limited, which makes it difficult to ensure an adequate income (#6). Some shops' organisation and management structures have also put pressure on their owners. In the most common case, friends or family members make the decision to launch a store, only later discovering the disadvantages of entrepreneurship, as partners faced with difficult business decisions. This has led to a rapid turnover of shops in Zurich and Geneva, with some spaces ceasing trading after a few months.[13] Reductions in store staff can cause other problems. Juggling customer interaction, telephone calls, and repair work requires constant task-switching; something in tension with the careful, sustained concentration required for complex repair operations. These cross-cutting pressures also impede employees' ability to estimate the duration of repair jobs, or calm users' anxieties and irritation.

Furthermore, the repair work undertaken in these spaces is structured by different repair regimes (Denis and Pontille, 2015). Outcomes can vary. Repairers' interventions might not restore smartphones to their original factory state—some interventions might tweak or damage the device, changing its performance and utility. Some repairers can even improve functionality, by upgrading a device's memory,[14] or boosting the battery's performance[15] (#4). They might make the technical changes required to "overclock" a device, boosting the processor's speed by installing a particular application or modified operating system[16] (#8). These examples show that maintenance and repair operations

13 This situation has lead to the appearance of simpler stores such as supermarket kiosks, which compensate for the high cost of rent by shrinking their size and simplifying their activities.
14 With external accessories and sometimes more hazardous DIY in the device itself.
15 By placing it on a Kaisi 9201 Battery Tester Battery Charger and Activation Board, originally designed for testing the voltage of iOS devices, which sends electrical power for ten minutes.
16 While this manoeuvre makes it possible to have a slower device, and therefore to be able to use several applications more efficiently simultaneously, it can make it less stable, generate more heat and therefore consume more energy.

need not reify an artifact's stability.[17] The coexistence of multiple repair and maintenance regimes is important, as not all repairs can be completed. To take one example, repairers may be unable to restore smartphones that have fallen in the toilet to an original baseline of functionality—even when dry, the battery may be permanently damaged, limiting the device's operating life. Incomplete repair jobs can, however, attract complaints from dissatisfied users. Even if repairers appreciate these different repair regimes and their limitations ("you can't always perform miracles"), this is not always appreciated by customers, who might attribute failure to wrongdoing or foul play (#3).

Another source of friction is the varying legality of interventions and repair practices. In extreme cases, customers make requests that would break the law, asking repairers to install a location-tracking chip, or recover data from stolen devices, but these jobs are usually rebuffed by store technicians. On a more mundane level, our portraits show how many repair jobs operate at the edge of what is permissible. Government regulations put legal restrictions on repair operations, while manufacturers go to great lengths to block the use of components from third-party suppliers. Depending on the repairer, and their level of comfort with informality, these boundaries can easily be crossed. This is the case *dessimlockage*, where the repairer intervenes to remove the limits locking (*simlocké*) a device to a specific mobile contract provider, allowing the use of any SIM card.[18] A similar service is jailbreaking, a way of circumventing software restrictions imposed by manufacturers, which allows new features to be added to the smartphone: changing the theme with new icons or different colours, or even adding pirated programs. In some repair venues, overclocking a device was classified as a kind of informal manipulation, while others seemed happy to install unauthorised programs or accessories.[19]

17 For more examples, see Marianne de Laet and Annemarie Mol's paper on the Zimbabwe Bush Pump, which describes the multiple ways in which villagers perpetuate this artifact by reconfiguring its boundaries and working order (de Laet and Mol, 2000); a similar phenomenon can be found in Lara Houston's work on mobile phone repair in Kampala (Houston, 2018), or Marisa Cohn's account of an ageing satellite, whose software needs to be frequently updated and adapted. (Cohn, 2019)

18 Devices marketed directly by operators as part of a package are subject to a "simlockage" that prevents users from passing over a competing network without first entering a deactivation code. This stems from the fact that most phones sold with a mobile plan have a 12 or 24-month commitment that forces the user to use only the operator's GSM, 3G or 4G telephone network.

19 One example of a (legal) app sometimes mentioned is "DU Speed Booster" for Android, which is described as "an optimization tool that allows you to increase the processing speed of your

Issues around questions of legality and informality are reinforced by the fact that customers are often unclear about what is or is not legal (#14). Perceptions and understandings of legality are mediated by users' backgrounds, habits, and the fact that certain regulations, such as SIM locking, have survived longer in some territories than others. With ready access to used and second-hand smartphones, and the skills to reconfigure them or access deleted files and content, repair shops are sometimes visited by the police. However, the authorities are more interested in the traffic and circulation of stolen electronic devices than informal repair practices or possible data theft.

Despite the precarity and informality of these spaces, most of our interlocutors affirmed their positive social dimension (#3 #5 #6 #7). If Fab labs and hackerspaces' social contribution is more obvious, as in, for example, "repair cafés," this is also true of most shops.[20] Customers come to drink coffee, chat about their devices, or seek help with minor problems (restoring settings, configuration), administrative services (such as filling on-line taxes forms), or a phone call in a language they cannot speak (#3). In these interactions, repair shops' fulfil roles far beyond their stated function, sustaining neighbourhood life and providing a social resource for the local community.

ENSKILLMENT

The range and diversity of smartphone repair techniques raises questions about the practical knowledge underpinning these stores' services, and how repairers' skills are developed and reproduced. Device manufacturers rarely provide resources or support for repairs made by users or uncertified third parties.[21] Critical information and specifications are kept private, with product designs, repair processes, and access to tools, software libraries, and official spare parts strictly limited to authorised repairers.[22]

Android device and clean the internal memory." https://du-speed-booster.en.uptodown.com/android

20 In this regard, hackerspaces and Fab labs may recommend certain repair shops for interventions that their repairers cannot carry out (#2 #10), contributing to local network formation.

21 Unlike computer security conferences in which hackers and industry specialists meet and exchange information in what Korn and Wagenknecht called an "arena of repair," i.e. "a discursive space that cuts across the social worlds of hacking and industry" (Korn and Wagenknecht, 2017, p. 2476)

22 Authorised repairers are the one who applied for certificates issued by manufacturers, which, in turn grant them with permissions to address a well-defined set of breakdowns and problems, following a specific procedure.

Our interlocutors complained about the existence of multiple smartphone models, and the technology's instability—with a smartphone's performance often contingent on some combination of updates, installed apps, the device's age, and previous repairs. Manufacturers have also enacted strategies to make repair more difficult. To take on example, both Apple and Chinese manufacturer (Huawei) have adopted non-standard, five-lobed security screws in their designs, to prevent people from opening their devices.[23] In another case, now infamous among users, an error message ("Error 53") appeared on iPhone models repaired with spare parts not sanctioned by Apple.[24]

Repairers use a wide range of information, knowledge and methods in their daily work, from troubleshooting to implementing solutions, identifying components or software solutions, and interpersonal skills such as listening or providing advice. How they developed their skills varied across our research sites, with individual repairers mobilising different practices, activities, and attitudes to build expertise and gain competence in their professional activities. While some got diplomas (#1), others learnt at home, alongside their family (#16).

An investigative approach to technical objects is common to shop owners and their technicians, as well as hackerspace founders and members. Deriving from a curiosity about technologies and how they work—something often associated with the figure of the hacker[25] (cf. Levy, 1984)—this often resulted in our interlocutors dismantling, tinkering with, and reassembling all kinds of devices. As one repairer explained, "I have always tried to understand things, to improve and to optimise everything" (#12). This attitude is central to the activities of hackerspaces and Fab labs, which frequently set out to teach participants how to "overcome the fear of opening an object" (#10).[26]

23 Until a company released a screwdriver specially for unscrewing iPhones, a development that demonstrates the co-evolution of large industrial groups and smaller players, with the latter constantly looking for ways to reassert control over technical objects.

24 While this outcome has since been abandoned by Apple, a 2017 software update (iOS 11.0.3) has proven capable of disabling certain features if repairs make use of unapproved spare parts.

25 And closely tracking what Pekka Himanen describes in his book on the "hacker ethic" (Himanen, 2001).

26 For more on this topic, see Berrebi-Hoffmann, Bureau and Lallement (2018, p. 151).

This investigative approach reflects a mode of "exploratory" engagement frequently associated with digital practices. Requiring a blend of effort, enthusiasm, and suspended attention, this mode of engagement was often applied to solve puzzles or technical problems without clear solutions (cf. AURAY, 2017).

Although most of our interlocutors had experienced this mode of engagement, recognising it as something they had often developed during childhood or adolescence,[27] this was not the case for everyone. Some people became repairers after seeing a business opportunity (#3 #5 #6) or having found work with on-the-job training (#3 #7). Many of these newer entrants have subsequently cultivated this investigative attitude. However, not everyone we met was necessarily inclined to pursue such an approach, with most repair tasks requiring the application of proven and well-codified techniques (changing screens, batteries, or buttons).

On questions of skills and training, our repairers' biographies cover a range of experiences, from formal technical instruction to pure self-study, often without a clear purpose. Though this may follow from the above ethos, this is not always the case. As our interlocutors noted, self-directed study is an essential part of the job, enabling repairers to handle the rapid succession of new phone models, as new components lead to new problems, and repair attempts are stymied by a lack of official resources or manufacturer guidance. In this, repairers continually refresh their skills, making use of online resources including YouTube video tutorials, web forums,[28] documents in pdf format, or membership of Slack or WhatsApp discussion groups.[29]

Among those interlocutors who received formal training, a wide range of qualifications had led them to the repair profession. Electrical engineering, computing, boiler-making, or communication systems were well-represented among founders and

27 An activity they often maintained, for some, by friendly/family sociabilities (#16)
28 While iFixit, GSM forum, cellphoneforum and sometimes Reddit are the most frequently mentioned, repairers also use forums in languages other than English, such as French, Arabic, Spanish, or Russian; a diversity that reflects both their personal biographies and the availability of online information.
29 The bulk of these resources concern specific intervention modalities linking smartphone models and a specific type of problem. But there are also comments on the repairability of certain devices, on new tools and machines, as well as advice on finding spare parts.

shop technicians,[30] while repairers operating out of hackerspaces and Fab labs had often had an academic background, with many holding engineering degrees.

What do repairers learn? Firstly, our portraits sometimes featured descriptions of certain procedures to be followed, as in the case of transferring a customer's address book (#1), cleaning a motherboard (#12) or diagnosing a problem with the device's camera (#16). In each case, repairers followed a clear operational sequence (*chaîne opératoire*)[31] imparted by their training. It was the practical application of such sequences that most interested us, revealing the repairers' bodily commitment and techniques—something that applies equally to diagnoses of failure, and the practice of repair. In each case, repairers feel, touch, push, pull, pull, move, blow, shake, and rub the devices[32] The repairer uses all of their senses, supplementing vision (detecting damage) with touch (testing a button, checking the integrity of a screen), hearing (noting a particular noise or its absence), or even smell (sniffing for strange odours during a micro-welding operation). In this sense, repair is an embodied and "sensual" practice (Dant, 2009, p. 16), even as repairer technicians' bodily capabilities are extended by instruments assisting in diagnosis (microscope, oscilloscope, battery tester) and repair tools (soldering iron, drying machine, reconditioning machine).[33] Ultimately, it is the repairer's body—the fingers that hold the device, the nails that pry open the case—that is their first recourse, and primary instrument.[34] Supplementing these bodily engagements, a secondary skillset, usually more recently acquired, encompasses the "soft skills" that extend beyond repair; listening to customers and their problems,[35]

30 Some of our interviewees displayed their diploma in store (#1).

31 "Chaîne opératoire" is a concept from anthropology, denoting operational sequences of gestures mobilised "to understand the propensity of human beings to associate material actions and physical objects with the production and rendering visible of social relations." (Lemonnier, 1992, p. 19).

32 A set of moves that seasoned repairers performed with greater skill, speed, and control than when one of the authors attempted the same tasks during their internship.

33 Such a commitment of the senses, particularly in the diagnostic phase, could be compared with the notion of prises proposed by Bessy and Chateauraynaud, which refers "to the encounter between a set of categories and material properties, identifiable by the common (supposed) senses or by instruments of objectivation." (Bessy & Chataurayanud, 2010. pp. 689–711)

34 "The body is man's first and most natural instrument. Or, more precisely, without speaking of instrument, the first and most natural technical object, and at the same time, technical means, of man is his body" wrote Marcel Mauss in 1935.

35 One of the repairers in shop #3: "Sometimes you are confronted with people who explain to you that their whole life is in their phones. I had a man once who wanted to get the pictures from his father's phone because it was all he had left of him, so it was really a life story."

offering advice, handling complaints, setting and negotiating prices. These situations require skills that far exceed any investigative attitude or formal training, with some respondents expressing their frustration at having to juggle repair jobs and their customers' social and emotional needs (#9).

These "soft skills" are also applied to questions of strategy, with shop managers, repairers and hackerspace participants constantly looking to develop their spaces, service offerings, and management techniques.[36] For stores, there is an additional commercial edge in staying up to date with new smartphone models and repair techniques: "As soon a new model comes out, we have to update our knowledge. We bought an iPhone 7, and our technicians broke it. It cost us 1000 CHF, but it was an opportunity to learn." (#12) Besides, since manufacturers rarely share the technical documents and specifications for these new models, individual repairers have had to take a more tactical approach to learning; dismantling and reassembling the new models—sometimes breaking test devices in the process. Among our repairers, this reconstruction of authorised repairers' operational sequences follows the kind of investigative engagement introduced above, driven by both their own curiosity and a need to adapt to changing customer demands.[37]

From our profiles, we identified three different ways repairers have tried to adapt to these changes:

- Reverse engineering, or discovery by trial and error, in which repairers try to fix their own smartphones or devices they have in stock. Some repairers document their attempts at this kind of reverse engineering, either via an online information system (#8), or, more commonly, in a notebook or paper binder that the repairer keeps updated (#4).
- Cumulative attempts at repair, experimenting on customers' devices over longer spans of time. Repairers encounter new problems or unfamiliar smartphone models as part of the normal flow of customer activity. Stores offer services that exceed

36 For stores, the search for additional income and revenue streams is a constant. Fab labs and hackerspaces experience a similar pressure in their push to involve a wider audience in their activities.

37 This pressure to acquire new knowledge and skills is sometimes linked to fears that the phone repair shop might not survive as a viable business model. During our investigation, several repairers reported that shops and small independents had ceased trading, deterred by the difficulty of handling problems beyond screen or battery replacements, and the barriers to repair put in place by device manufacturers.

their expertise, using these encounters to develop their business, and explore new possibilities. This relies on a form of improvisation and exchange between colleagues.[38]

- The authorship and use of documents, explanatory videos, or contributions to discussions on online forums. This includes "teardown reports" produced by commercial third parties such as Chipworks or TechInsights,[39] who dismantle smartphones to produce lists or plans detailing a device's technical specifications and identified components (processors, sensors, antennas, screens, etc.). Although these documents can be expensive, running to several thousand euros, copies circulate among personal networks, while some repairers locate incomplete or bootlegged versions on unofficial websites. Not all of their contents are of practical use, but we found screenshots and annotated photographs among some of our respondents' notes (#4).

Not all repairers are able to set aside time for research activities based on documents that are difficult to access, and problems rarely encountered by their customers. Hackerspace members may see an intrinsic value in exploratory tinkering, but most commercial repairers lack the time, motivation, or ability to engage. As a result, different repairers develop varying levels of skill and expertise, based on their exposure to different problems and phone models. Unable to provide a certain skill or service, some repairers might direct their customer to a more specialised competitor (#2). Equally, exposure to a certain set of problems—and identifying their commercial potential—can lead a repairer to gradually specialise in a specific area, without having planned to do so. This was the case for two of our respondents; one launched a company focused on screen reconditioning (#5), while another became a nomadic worker, without a store home base, specialising in micro-welding (#4).[40]

38 Paralleling Julian Orr's observation that repair is not just an application of existing knowledge, but turns on the relations between customers' problems, the condition of machines, and the actions of technicians themselves (Orr, 1996, p. 2)

39 Two companies that merged during the final stages of our research. Based in Ottawa, the new firm describes itself as "Your partner for revealing innovation and analyzing patent rights." See https://www.techinsights.com/about-techinsights.

40 Beyond this specific case, during the 2020 COVID-19 lockdown, most of the shops turned their businesses into nomadic operations, while some asked their customers to pre-book appointments over the phone

Although we have stressed the importance of individual repairers' skills, knowledge and experiences, it is important to recognise the material elements of their know-how, which frequently corresponds to their documentation of certain techniques, and the acquisition and use of tools and spare parts.

Our interlocutors commonly used notebooks full of drawings, technical diagrams, and lists of components with description of their characteristics and shortcomings. They also assembled binders of extracts from forums and annotated screenshots of web forum posts, sometimes in multiple languages (#4), books to describe maintenance procedures (#5), video series (#3), and even an entire "internal learning system" compiling failures and notes on the availability of spare parts (#6 #8). Rarely intended for circulation beyond a group of colleagues and trusted contacts from other stores, these documents may belong to an individual repairer (#4), a shop (#3 #5), or an entire chain (#8). More detailed or specific supplementary information may be shared with the same small circles on platforms such as WhatsApp (#12), on online forums, or through face-to-face meetings (#6). Unlike hackerspace participants, the communities of practice formed by commercial repairers exists in more limited and transitory ways, with restrictions on who is able to benefit from resources and shared information.

Alongside these examples of documentation, we can see the material components of repairers' skills and expertise in their maintenance of a stock of spare parts and components, and the customisation and use of repair tools. Such practices were relatively common among our respondents, with many repairers having cultivated extensive collections of spare parts, but also fully-functional smartphones, earlier mobile phones (#1), and other electronic devices such as tablets or smart watches (#6). These devices and components are usually distributed across multiple boxes, cabinets and drawers, typically in the back of the shop, away from customers. Representing a stock of materials easily accessed during a repair job, individual items constitute what David Edgerton has described as "reserve technologies," technical objects used as temporary or short-term solutions to a problem (Edgerton, 2006, p. 52). By pulling materials from this reserve, a repairer can replace defective smartphone components with their counterparts from other, similar devices—a creative reuse of technology described elsewhere as "cannibalization" (Callen 2017; Houston 2017). This stretches notions of creative repurposing, a practice close to what Claude Lévi-Strauss saw in the "bricoleur,"[41] which encompass the reuse and improvisational adaptation of spare parts from other devices, and the use of tools that may not have been intended for a particular job (Jackson et al. 2014, p. 7). In one case, we saw a microscope improved with a surplus phone case and LEDs placed to provide better lighting (#4). The repairer who made this device, a nomadic worker specialized in micro-soldering, explained how he had needed such a feature, but could not afford to buy an expensive microscope. Re-using spare parts to improve an existing tool was not only a way to save money, but also helped him show colleagues his ability to repair and modify complex technical objects.

This issue of spare parts, and who has access to them, highlights another important aspect of repairers' work: their presence and participation in multiple networks and communities of practice. Although repair work, like our informants' acquisition of skills and know-how, is materially situated, it is also a collective effort.

LOCAL AND GLOBAL NETWORKS

In focusing on portraits and testimonies of individual repairers based in shops, hackerspaces and Fab labs, our account might make it seem like maintenance and repair are straightforward endeavours, requiring the participation of just a repairer and a customer. If we were to limit our analysis to repairers' workshops, and repairer-customer interactions, we would miss how diagnosis, problem-solving, and repair work draw on broader networks of people, documents, and components. To avoid this trap, we apply a more structural analytical lens to the work of maintenance and repair, situating individual repairers' interventions in a wider field.

These networks connect repairers with a variety of suppliers and subcontractors, spanning different roles in the supply chain, and capable of manufacturing and distributing key components. Such networks are an important part of repairers' capacity to intervene. Without access to their resources, it is difficult to replace

41 The bricoleur is a figure for whom "the rules of his game are always to make do with "whatever is at hand," that is to say with a set of tools and materials which is always finite and is also heterogeneous because what its contain bears no relation to the current project, or indeed to any particular project, but is the contingent result of all the occasions there have been to renew or enrich the stock or to maintain it with the remains of previous constructions or destructions." (Levi-Strauss, 1966, p. 20).

broken or faulty components such as motherboards or connectors. Sourcing replacement parts can be a problem for hackerspaces and Fab labs (#10 #11). A lack of stock can leave them unable to make repairs, or relying on participants to supply their own parts (#11). Even stores and businesses may struggle to access quality parts at a non-prohibitive price (#1 #5 #6). Although it is common for repairers to order components directly from Chinese online platforms, such as Alibaba, our informants also used official suppliers (when they agree to sell parts) and European resellers and intermediaries (e.g. #1, who mentioned his suppliers in Paris). They may have difficulty in controlling for quality, sometimes receiving components that prove unusable. As a result, a small number of repairers have, in the past ten years, started "going to the source," visiting shopping centres in Huaqiangbei and Shenzhen to buy parts from local manufacturers and suppliers (#17). Some of our interlocutors underlined the strong relationship they have built over time with particular Chinese suppliers;[42] a source of competitive advantage that allowed one French store to reposition itself as a spare parts supplier, serving other repair businesses (#7). Parts sourced in Huaqiangbei are sometimes not always what they claim to be, however, and may be manufactured by companies reproducing components, or supplied by craftsmen specialising in disassembling second-hand smartphones.[43]

Alongside spare parts, the acquisition and use of more complex tools, computer programs, and firmware libraries require repairers to have cultivated a network of suppliers. This topic rarely came up during our interviews and observation sessions. Although repairers were reluctant to expand on their use of such tools in their work, any adoption of specialist apparatus usually followed recommendations from our interlocutors' ex-colleagues and informal contacts, or via specialist online forums.

Repairers are also participants in a second type of network, a more knowledge-focused community of practice oriented around the repair and maintenance of digital objects.[44] Beyond

42 Whose identity they did not want to disclose.

43 See Yvan Schulz (2017, p. 85), who, in his investigation on re-assemblers of this "hardware mecca," describes how "tens of thousands of craftsmen and merchants dismantle, clean, sort, test, assemble, display in shop windows and online, negotiate, pack, [and] send" spare parts.

44 A "community of practice" is a group of people whose identity is defined by their common interests and competences, their engagement in discussions and joint activities, and sharing of information, tools and know-how (Lave and Wenger, 1991; Wenger, 1998).

individual practitioners, this network also includes firms specialising in particular services, organisations like iFixit that produce their own documents and videos, research offices publishing teardown reports and schematics, and active members of online forums. Such groups also rely on unplanned meetings between repairers from different stores (#4 #6), exchanges on local WhatsApp groups (repairers in the vicinity of Lake Geneva, in the case of #6), or transnational meetings with colleagues or distant cousins (#4 #16). Information on repair techniques, technical issues,[45] and phone components[46] is made available, and exchanged all day long between repairers. Although information shared on forums is easily found, it is harder to locate detailed technical documents or first-hand summaries of individual repairers' reverse engineering efforts; something that often relies on mutual aid between people who are socially close, if not living in the same neighbourhood. This collective aspect of repair also features in Lara Houston's research on mobile phone maintenance in Uganda, a setting where repair practices were "shaped by connections to trans-local bodies of repair knowledge, largely accessible via the Internet" (Houston, 2019, p. 130). For Houston's respondents, a lack of internet access limited her subjects' attempts at repair, but this was not the case for our interlocutors—in part, because their interventions were based on learned processes (repair of damaged parts). Remote access was only needed for serious problems (on hardware layers such as the motherboard, or software issues) or when handling new smartphone models.

The number and range of actors in these networks show it is difficult to work with smartphones as a lone repairer, without active, regular participation in a wider community of practice. Building their networks, repairers must cultivate relationships with trusted component suppliers, while knowing where to direct customers with problems beyond their expertise.

Probing these networks, we find a dynamic that fluctuates between competition and cooperation. In our observations, repairers moved between stores—either physically or via digital media (#1 #4 #6)—to access specific tools or parts, but also to seek or provide

45 These exchanges reflect Julian Orr's description of the collaborative nature of photocopy repair, where discussions of common problems, causes of failure, or common exceptions become stories that circulate among repairers (Orr, 1996, p. 127).

46 To take one example, the repairer profiled in portrait #4 showed us binders filled with macro photographs of motherboards.

advice and support. Offering time, guidance, or resources may help to solve customers' problems, but it also contributes to the upkeep of a community of practice. However, where hackerspaces and Fab labs were designed to facilitate participation and reciprocity, our observations show the commercial repair ecosystem is more complex, comprising a diverse range of actors, with multiplex ties connecting repair stores, spare parts suppliers, resellers, intermediaries, and subcontractors.

Networks established by these repair businesses are shaped by multiscale interactions and a sustained process of "globalization from below." The first of these concerns the interplay of local, international and global networks. Repair stores in Geneva, Zurich, and Lausanne work with partners and subcontractors in their home city, but also interact with European intermediaries—particularly in France or England—as a way to avoid expensive shipping costs. They also negotiate with contacts in Shenzhen, Hong Kong, and Dubai for access to schematics and spare parts. In addition to providing an operational base for individual repairers, the repair shop thus also serves as an underappreciated node in a globally distributed supply chain. Making the most of these transnational networks requires skills beyond the reach of some repairers, however—most commonly language competencies,[47] but also negotiation skills, awareness of broker and dealer networks, and the confidence to initiate contact (#6 #7 #17). This could explain why some shops seem better able to handle diverse technical issues, establish and maintain a local reputation, or work with a range of different smartphone brands. The quality of a shop derives, in part, from its employees' ability to mediate between local and global scales. Locating a specific component might involve a chain of actors from Geneva to Shenzhen. Similarly, rapidly replacing a colleague or fired employee could require calling on a contact in the same city or reaching out to a cousin overseas.[48] In these cases, the most resourceful repairers or shopkeepers are often those best able to leverage a varied and geographically-extensive network. This form of cosmopolitanism[49] is particularly common among repairers from

47 French and English are most common, but also Arabic, Thai, and others, which offer access to different networks.

48 Sometimes with all the difficulties that this implies in terms of obtaining work permits.

49 Close to Alain Tarrius' notion of "circulatory territory," as described in his work on Mediterranean migrants communities: "A certain type of socialisation which is proper to places that are paths for population transit. Individuals recognise themselves inside the places they occupy or through which they pass, in a common history of mobility that induces an original social bond." (Tarrius, 2005) the French Maghreb (#6 #7) and Arabic (#4 #16) diasporas who have an ability to code-switch between languages and registers, calling on extended family networks, and travelling to establish relationships with international suppliers.

A second factor shaping smartphone repairers' networks and communities is their implication in processes of "globalisation from below" (Portes 1997). Anthropologist Gordon Mathews has recently defined this as comprising "the transnational flow of people and goods involving relatively small amounts of capital and informal, often semi-legal or illegal transactions"—something often associated with "the developing world' but in fact apparent across the globe" (Mathews et al., 2012, p. 1). Commercial repair shops are a space where this phenomenon is readily apparent, particularly among those with staff and owners from international or diasporic communities. Although the repair technicians who work in these stores are not truly nomadic, they are active participants in those networks constituting and facilitating "globalisation from below." At the same time, they also trade in high-tech goods and services, a sector with its own norms and values, and which they must navigate to promote accessories (chargers, headsets, speakers), buy smartphones, tablets, or displays, or seek advice from those selling the most popular brands. However, not all repairers follow of this double logic of multiscale interactions and "globalization from below" (#2 #8). The skills in navigating such networks are unevenly distributed among our respondents, and it seems that shop founders in Switzerland are rarely involved in such dynamics. Operating beyond the reach of the dominant smartphone brands, these repairers must juggle the local and the global, mediating between the "bottom-up," ad hoc logics of informal networks and the structuring influence of "top-down" manufacturers—much like the Huaqiangbei repairers who sell refurbished smartphones, as described in Yvan Schulz's research on Shenzhen (Schulz, 2017, p. 87).

Shaped by the established digital economy, these networks' dual logics—straddling the global and the local, the "top-down" and the "bottom-up"—underpins an alternative model of innovation, which departs from the technology-led model favoured by digital start-ups and the major brands and manufacturers.

FROM REPAIR TO SILENT INNOVATION

Contrary to some customers' expectations, repair and maintenance practices do not always return a smartphone to its original state. Concurrent and overlapping regimes of maintenance see repairers making "improvements" that go beyond resolving a clear or specific problem. These technical innovations might involve changing a smartphone's appearance (#5), overclocking the processors (#8), boosting the battery (#4), or installing apps to optimise the device's speed.[50] These interventions may boost a smartphone's performance or unlock new capabilities, but they are risky, and can cause lasting damage. Such interventions show how technicians can reconfigure a given device's functionality, even when facing clear technical and resource constraints.

Capitalising on technicians' exposure to a range of customer requests, a second form of innovation can be seen in repair stores' organisational strategies and commercial offerings. Posters promoting services, pricing, and supported phone models are regularly replaced, to reflect a constantly shifting business offer. Though most repairers start out focusing on hardware—typically broken screens and buttons—and minor software problems, we witnessed a proliferation of techniques, as new hardware problems required a range of careful interventions. Software issues proved similarly varied, as users sought help to optimise or reconfigure their devices—something that could involve adjustments to virtual assistant software, recommendation algorithms, or apps designed to extend a device's battery life. This proliferation of services tracks the smartphone's growing complexity, but this cannot account for the full extent of repair stores' diversification. Instead, as many of our respondents explained, customer requests led them to rework of improve their services. Faced with shifts in demand, technicians found new ways to develop, document, and share their know-how—implementing organisational innovations that have, in turn, expanded their repertoire.

Running alongside these new repair services, we also saw a series of social innovations in how repairers supported their customers' smartphone usage. These channels of interaction saw repairers providing advice and recommendations, both face-to-face and by distributing documents codifying best practice. Some repairers focused on providing users with tips and tricks in-person. In some cases, this involved sharing techniques to coax extra capabilities

50 By using, for instance, "DU Speed Booster" for Android.

from an old or broken smartphone—deactivating the device's GPS to maximise its battery life, disabling app notifications,[51] or learning to create regular backups "in case the phone falls into the toilet." Some stores distributed printed documents codifying these techniques, with step-by-step instructions offering greater detail. In one case, we saw a flyer showing different ways to extend a battery's lifespan, with detailed advice and explanations of the common causes of certain technical problems. As well as helping individuals, these documents also served as a form of marketing, promoting the store, and validating smartphone users' concern for the care and longevity of their devices.

In some cases, this kind of help extends beyond the smartphone, addressing other areas of everyday life impacted by digital technologies. We have, for example, seen people come to repairers for help with administrative tasks, such as making a doctor's appointment or filing their tax return online (#4). Although fab lab and hackerspace repair cafés were equipped to provide help and guidance with smartphone use, this support was more visible in commercial shops, particularly those with strong links to working class and minority ethnic communities. Just as copy shops once played the role of public writer[52]—and still do, in some places—our repairers sometimes provided informal instruction to customers struggling with contemporary technology. In this, many repair stores have cultivated "digital literacy,"[53] playing a role that is crucial to their customers, but easy to overlook.

In many ways, this educational function has fulfilled many of the promises and public expectations of Fab labs and hackerspaces. Over the past 15 years, many public organisations claimed such spaces would contribute to new kinds of design and manufacturing, serving to "democratise innovation" (Anderson, 2012), and symbolising a break with established models of Fordist mass production. It was hoped they would help participating users to overcome their lack of technical skills, contributing to the "opening" and reconfiguration of technical objects and consumer devices (Liefooghe, 2018). Our research found that although public

51 Reducing the load on both the battery and the user's "mental health," as one salesman described it.

52 A public writer provides support for people wanting to write a letter, for administrative purposes or otherwise.

53 A term the OECD (2000) defines as the ability to understand and use digital technologies, and to participate in a society where they are increasingly important across multiple contexts of everyday life.

hopes and expectations for these spaces were not wrong, they were clearly overhyped—with the activities and interactions anticipated by these early visions occurring just as often in repair shops as in Fab labs or hackerspaces.

In considering how repair and maintenance facilities contribute to innovation, our portraits show how repairers have handled technological change (and the increasing complexity of smartphones), responding by diversifying their services and, increasingly, supporting users in reconfiguring and customising their digital technologies. This combination provides an alternative to dominant models of innovation, which foreground the role of technological R&D in generating intellectual property that then can be applied in new products and services. Although our informants' activities share certain characteristics with forms of bottom-up innovation (von Hippel, 1988), we saw little evidence of repairers or stores making efforts to commercialise products based on the results of their activities. At the same time, the three modes of innovation we encountered diverge from descriptions of inverted or frugal innovation (Radjou & Prabhu, 2015), as found in the developing economies of the Global South. Instead, we characterise these stores and repairers' innovative practices as a form of "silent innovation."

In this, repair shops are one example of the kind of organisations that emerge where the networks facilitating processes of "globalisation from below" intersect with formal, regularised commercial activity. As an upshot of their repair practices, these repair businesses generate and circulate new techniques, practices, and service offerings. Such "silent innovation," we suggest, comprises three components:

- The accumulation of situated knowledge through on-the-job learning, a result of sustained interactions with customers and their specific problems;
- Practices of reverse engineering and exploratory material engagement;
- The activation of weak ties within a community of repairers, a professional network incorporating local and global elements.

Perhaps unexpectedly, those repairers we spoke to expressed doubt that their practices represented any kind of "innovation"[54].

54 "No. We do not do innovation here! We will extend the life of the device, but we are not doing miracles! An S4 Galaxy will not become an iPhone 7 after a repair," as one manager told us.

"MY STORE IS A LABORATORY"[55]

Initiatives to reduce the environmental impacts of digital technology are increasingly common. From municipal incentives to corporate social responsibility programmes and proposals for international regulation, questions of maintenance and repair are a growing concern.

Public debate and press coverage of these issues has often focused on the maker movement.[56] Initiatives such as repair cafés were invested with the high hopes and expectations of those advocating for a change in how people relate to technical objects. Some claim the social dynamics of Fab labs and hackerspaces have empowered users to re-appropriate their technology, with greater access to machines and equipment helping to extend the lifespan of consumer technologies.[57] Garnering less public scrutiny, the rapid spread of repair stores in the past 10 years has run alongside these more high-profile efforts to democratise access to technology. With small repair businesses achieving many of the same results as those pursued by Fab labs and hackerspaces, the mainstreaming of less "charismatic" forms of maintenance and repair reflect users' growing concern for the sustainability of everyday technologies.

By comparing different repairers' practices and life courses, our research shows how the maker movement's stated goal of "democratising" technology has proven more challenging than many expected. The Fab labs and hackerspaces we visited are distinct spaces, providing a clear alternative to dominant forms of technology, commerce, and innovation. Supporting participation and citizen empowerment, they provide space and resources for projects that would not have been possible without institutional support.[58] Engaging different audiences in the production and re-appropriation of technology through a range of events and programming, these institutions provided a dedicated space for reflection, discussion, and the interrogation of technical objects. Although repair café events were well-placed to extend the lifespan of

55 The words of a repair shop founder not included in our portraits; based in Geneva, he argued that his store constantly buys new smartphones in order to understand how they differ from earlier models.

56 See Lockton (2013), Fairware (2016), Ruskin (2016).

57 See Gershenfeld (2007); Anderson (2012); Cavalcanti (2013), as well as Braybrooke and Smith (2018) for a description of this kind of citizen empowerment.

58 Beyond the examples provided in our portraits, see Bosqué (2016), and Berrebi-Hoffmann, Bureau and Lallement (2018) for detailed descriptions of similar such contributions.

everyday objects such as toasters and vacuum cleaners,[59] the successful repair of smartphones has proved elusive, seemingly beyond the capabilities of those involved. Despite abundant online resources and participants' demonstrable technical expertise, the repair of more complex digital technologies seems contingent on other factors. These include a given repairer's mastery of hardware components and specialist software, their active participation in a wider community of practice, and access to a regularly-updated stock of spare parts and quality tools—skills and resources easily located in the repair store environment.

Although Fab labs and hackerspaces rarely acknowledge maintenance and repair in their stated objectives, they have cultivated a distinct attitude to technology among their user base, helping people overcome a fear of opening and tinkering with their own devices. Those hackerspace members who tried to engage in phone repair encountered significant difficulties, however, facing limited access to smartphone parts, lack of specialist tools, and problems beyond their skill level (#10 #11 #15). In this, we can see that it was not the stated aims or goals of the maker movement that limited their impact, but, instead, how such spaces were promoted and publicised. Overburdened by enthusiastic media coverage and public discourse, the early promise of these venues and events set expectations they were simply unable to meet.

Contrasting unfavourably with the portraits of repairers that comprise the bulk of this book, the obstacles encountered by those fab lab users trying to repair complex digital devices suggest the need for an alternative figure of maintenance and repair. Less appealing to supporting institutions than the celebratory discourses that recast users as "makers," the figure of the in-store repair technician now seems as relevant as any "empowered" hackerspace user, if not more so.[60]

59 As described in the many ethnographic studies of hackerspaces (e.g. Bosqué, 2016; Goyon, 2016).

60 Although the hackerspaces and Fab labs we visited struggled to repair smartphones and digital artifacts, they played an important role in the 2020 COVID-19 crisis, fabricating equipment for the pandemic response including masks, respirator valves, and face shields.

AN ALTERNATIVE FIGURE

The repairers we met in the repair shops of Geneva, Lausanne and Zurich have a commercial imperative to serve customers, finding ways to quickly solve their problems. Such an objective rests on their ability to ensure a positive outcome. Typically rooted in an exploratory ethos, a repairer's effectiveness requires efforts to develop their know-how, access to equipment, a stock of spare parts, a strong community of fellow repairers, and a network of suppliers and subcontractors. Their individual experiences are shaped by the range and frequency of repair jobs, and the pressure to adopt new models and innovations pioneered elsewhere. This, in turn, improves repairers' ability to serve their customers.

From these portraits and testimonials, it appears the maker movement's early promise—pursuing user empowerment in order to boost technical objects' durability and lifespan—has partly occurred elsewhere, in small repair stores, and at the repair technician's workstation.

It is, however, important not to simply replace the figure of the user-turned-maker with that of the endlessly resourceful commercial repairer. In centring the role of individual actors, the heroic figure of the lone repairer can mask the many systems and structures supporting their work. Such accounts might also be prone to sentimentalising maintenance and repair, eliding the real challenges and frictions of the role.

Repair shops experience high staff turnover, facing pressure to hire and fire employees in response to fluctuating levels of demand. The work is difficult, and the role of the repairer is often precarious. Repair capabilities shift and evolve as new phone models are released, leaving those in the profession racing to keep pace with a moving target. The required skillset is constantly changing, but with few formal institutions to support training or professional development. Furthermore, even where smartphone manufacturers and telcos have proven willing to concede the value of maintenance,[61] repair jobs have smaller profit margins than products or services, and resist scaling.[62] As our portraits

61 Sometimes even opening their own repair spaces, as in the case of Salt, a phone provider, which launched a "repair corner" as an in-store facility sometime around 2018.

62 In a business context, "scaling" refers to the ability of an organisation to sustain or improve its performance, in terms of efficiency or profitability, as demand increases. In services, it show, in-store repair operations can be complex, requiring great dexterity on the part of technicians. Depending on the issue, a job might be quickly resolved or else require more time and care, contingent on the purchase of an unavailable part, or the expertise of a busy subcontractor.

Demonstrating how these repair shops diverge from the organisational models and practices of other commercial businesses, our portraits of repair technicians show how skilled support provides phone users with new opportunities for users to circumvent the artificial restrictions imposed by large manufacturers, and thereby re-appropriate consumer technologies. Our interlocutors' testimonies revealed the extent to which they were able to realise many of the original aims of the maker movement. By providing users with the means to adjust or reconfigure their devices to better meet their needs, shop repairers have come to practice what Laurence Allard dubs "counter-making" (ALLARD, 2015, p. 164).

HIDDEN LABORATORIES OF CARE

The stores we visited provide a venue for maintenance and repair practices, while equipping repairers with the resources to negotiate the cross-cutting logics of commercial networks and their own communities of practice. These are experimental spaces, where repair technicians combine DIY practices and situated knowledge to reconfigure black-boxed technical objects. At the same time, these are also sites where knowledge accumulates about smartphones and shared with various networks. In this, repair shops are a kind of laboratory, a dedicated space for maintenance and repair, but also practices of innovation and knowledge production. The main differences between Fab labs and repair labs follow repairers' commercial focus on resolving their customers' hardware and software issues—a clear contrast with hackerspaces and fab lab's emphasis on "making."

In taking one interlocutor's claim that "my store is a laboratory" at face value, we can come to appreciate how this may be more than a simple metaphor or analogy. Noting the extensive body of literature on laboratories as sites of knowledge production,[63]

means the capability of a company to handle a growing amount of work, or its potential to accommodate growth.

63 See LATOUR & WOOLGAR (1979) for a seminal book in the laboratory study tradition within Science and Technology Studies, KNORR-CETINA (1999) for a discussion of the diversity of our portraits underline how phone repair stores, Fab labs, and hackerspaces share common features with the scientific laboratory. In working to open black-boxed research objects, repairers produce new knowledge[64]—something with value for the individual repairer, their customers, and members of their professional networks and wider communities of practice.

Considering the distinctive characteristics of these repair spaces, we identified three ways in which they differ from more familiar scientific laboratories: (1) generating knowledge through the disassembly and reverse engineering of existing technology; (2) taking an explicitly critical stance towards multinational manufacturers; and (3) combining practices of knowledge creation with an ethic of care.

First, the "knowledge" produced by repairers has a different epistemic foundation from that of their scientific or R&D counterparts. Although repairers' communities of practice may replicate the networked structure of earlier academic networks, their approach diverges from the social and institutional norms of scientific research and applied R&D. Repairers generate novel insights through practices of disassembly, reverse engineering, and trial-and-error exploration. A form of research-through-unmaking, this is a counter-R&D that aims to reproduce the ways smartphones are configured, but also provide the means of reworking or re-appropriating them or without the approval of their manufacturers. These practices mirror examples of "Critical Making," a term Matt Ratto and his colleagues use to describe knowledge generation grounded in creative practice, emerging through material engagement with technology (Ratto, 2011, p. 253). Extending this notion to our own research, an equivalent "Critical Unmaking" would centre the knowledge gained from dismantling technological objects—something common to all the repairers we observed, regardless of their institutional base. The critical difference between repair shops and hackerspaces, we argue, lies in the difference between systematic and exploratory processes of knowledge production. In the case of the phone repair store, efforts to produce usable knowledge are a source of competitive advantage, while Fab labs and hackerspaces' emphasis on

scientific cultures of knowing, Vinck (2003) for a perspective on design engineering and innovation, and Vinck (2007) for a more recent discussion of the notion of "laboratory."

64 A definition of laboratory activities proposed by Dominique Vinck, relying on ethnographies of research centres, as well as STS studies from the past 30 years (Vinck, 2007, p. 162).

participation has supported a more exploratory, open-ended mode of engagement.

The "critical" qualifier in "Critical Unmaking" highlights a second difference between the laboratory and the repair shop. These shops' repair activities are shaped by a broadly critical stance towards large manufacturers and multinational companies. Restricted by the ambiguous legal status of many of their operations, repair shops rely on a "technological disobedience"[65] in their strategies and actions. Often directly threatened by dominant consumer brands, and working against their efforts to prevent users from tinkering with their own devices, repair shops' work is a form of "counter-making," producing iterations of the smartphone which differ from those envisaged and sanctioned by manufacturers (Allard, 2015, p. 164). In this, these sites offer a counterpoint to the kinds of R&D producing the widely-hyped innovations that feed into new models of smartphones, connected objects, and digital services. Although the impacts of these shops are dwarfed by the major digital manufacturers' market-shaping power, repair shop laboratories support users' efforts to prolong their device's lifespan; finding ways to circumvent manufacturers' technical restrictions, and acting as a brake efforts to perpetuate a cycle of regular device upgrades.

Third, we saw how these "cell phone clinics" worked to reconcile knowledge creation with community engagement and an ethic of care. Having opened this book with a litany of shop names highlighting the different metaphors used to describe repair stores' activities, the frequency of medical analogies stands out as significant. These framings underline the similarities between our repair shop laboratories and clinical research. Regardless of the appropriateness of equating commercial repairers with medical researchers, these stores' techniques for diagnosing and documenting smartphone problems—experimenting with different interventions, deriving general lessons from specific cases, producing documents, and socialising new knowledge—demonstrate clear parallels with the tenets of clinical research.

Applying their exploratory ethos to disassembly and reverse engineering, adopting a clear stance opposing the power of big

65 Coined by Cuban artist Ernesto Oroza, this term designates the unorthodox and liberatory character of the small, humble gestures of collection, repair and reuse, and their potential to alter and question means of production (Oroza, 2009, p. 20).

business, and framing their work as something with a para-clinical dimension, repair shops a site of friction where different logics and epistemologies intersect.

A laboratory of sorts and site of friction, the repair shop could equally be cast as a "patch," situated in a wider landscape. This analogy builds on environmental anthropologist Anna Tsing's work, where the term "patch" names pieces of apparently derelict land, from which precarious workers are able make a living, by picking Matsutake mushrooms (cf. Tsing, 2015). Foragers in the forests of Oregon—and elsewhere—supply these highly prized mushrooms to independent buyers, who resell them to agents of global trading companies, in order to meet demand from customers in Japan, where the mushroom has fallen victim to changing agricultural practices. For Tsing, these "patches" represent the outermost edges of the capitalist system, spaces where more familiar and scalable forms of organisation flounder, incapable of extracting value (Tsing, 2015, p. 111).

Comparing the phone repair shops we researched with Tsing's "patches" reveals the limits in casting such spaces as a situated subtype of the scientific research laboratory. In the course of their repair activities, shop technicians must manage an interplay of multiple, often divergent logics. These stores are, at once, commercial service providers and sources of local aid and social support. Repairer technicians are active participants in high-tech, globalised supply chains, even as they mobilise low-tech and improvised DIY practices. Applying their skills and expertise for profit, these repairers nonetheless oppose the might of large manufacturers and, in certain cases, can be seen working for free. Just like Tsing's mushroom-pickers, our portraits show in-store repairers making their livelihoods at one end of a distributed global commodity chain. Acting with a level of autonomy, even when working for a chain or larger store, their everyday repair practices show how money can be made without "rationalizing labor and raw materials… [through] acts of translation" (Tsing, 2015, p. 98–100). Like other patches, these repair shops "emerge from disturbance" (ibid., p. 139), responding to demand unmet by the major smartphone manufacturers, who have disavowed repair as flatly unprofitable. Approximate, shifting spaces on the periphery of the consumer technology landscape, these repair stores are "neither neat solutions nor guarantees of progress" (Perng, 2019, p. 14), even as their position allows them to resist co-optation or scaling. Despite the webs of cross-connections sustaining a wider community of repairers, each repair store has its own history and culture, and the practice depends on the continued existence of larger manufacturing firms and, indeed, component suppliers. Individual repairers may be a source of innovation, devising new methods and practices to ensure their economic survival, but their room for growth has an upper limit. If they were able to scale further, phone repair stores would no longer be a source of value to the dominant manufacturers, but instead threaten their strategies of regular upgrades and planned obsolescence.

This book opened with an assumption that a close examination of the varied practices taking place in repair shops and hackerspaces could enrich our understanding of efforts to improve the durability of digital devices. The portraits shared here, our ethnographic vignettes, and subsequent analysis highlight the importance of these spaces for technological and social innovation. We also found that these venues did not necessarily hang together as a coherent unit, with striking differences between hackerspaces and commercial repairers. As a result, we narrowed our focus to repair technicians working in shops, and, more specifically, how these actors generated new knowledge about "closed" smartphones and consumer devices. Using analogies from the existing literature, we compared these repair shops with scientific research laboratories and "patches," spaces of disturbance and transformation situated within larger landscapes of technological practice. Although these repair shops are likely to play a larger role as societies face growing pressure to transition to a more sustainable economy, we conclude by asserting that their emergence was not a given, and that their continued existence should not be taken for granted. These observations, and our repairer portraits, prompt further questions for policymakers and practitioners. Should these venues receive greater support from government and public institutions? Do they have the potential to contribute in areas beyond those identified in our portraits and vignettes? If we are to rework our relationship with technology and consumer devices, is a greater focus on repair and maintenance enough? Is this just a question of fixing, or do we need new metaphors to imagine new—and less resource-intensive—interactions with smartphones, electronics, and other consumer devices?

#

1G/2G/2.75G (EDGE)/3G/4G/5G

Various generations of cellular network technology (communication networks the network is distributed over land areas called "cells," each served by at least one fixed-location transceiver, but more normally, three cell sites or base transceiver stations).

B

BLOATWARE

Unwanted software (included on a new computer or mobile device by the manufacturer) which may use large amounts of memory and ram, and eventually slow down the device.

BODGING

A clumsy or inelegant job, usually as a temporary repair (see also kludge).

C

CERTIFICATE (APPLE)

Apple certified ios technician (acit) is a four-day course that prepares participants to become certified in troubleshooting and repairing iphones. Acit 2019 qualifies a technician to repair ios products that were produced before march 15, 2019.

CPU (CENTRAL PROCESSING UNIT)

Sometimes referred to simply as the central processor, but more commonly called processor, the cpu is the core of the computer/smartphone where most calculations take place.

D

DATA CORRUPTION

Refers to errors in computer data that occur during writing, reading, storage, transmission, or processing, which introduce unintended changes to the original data.

May damage the quality of digital content. Can be solved with various countermeasures (data scrubbing, data redundancy) or via data recovery software designed to recover corrupted data like photos, audio file, videos, sms, call history, contacts, whatsapp messages and other files that's deleted or formatted.

DEFRAGGING

Short for "defragmentation," a technique to consolidate files on a hard drive. Smartphones as devices that use flash storage, which is not affected by file fragmentation like traditional hard drives are, do not need defragmentation. However, some repairers may defrag android devices only to wear out the storage and speed up the device, but it may eventually lessen its lifespan.

DESOLDERING

The removal of solder and components from a circuit board for troubleshooting, repair, replacement, and salvage.

E

E-WASTE

Discarded electrical or electronic devices. Used electronics which are destined for refurbishment, reuse, resale, salvage, recycling through material recovery, or disposal are also considered e-waste.

F

FAB LAB

A small-scale workshop offering (personal) digital fabrication machines. Set up to provide the environment, skills, advanced materials and technology to make things cheaply and quickly anywhere in the world, and to make this available on a local basis to entrepreneurs, students, artists, and small businesses.

FLASHING (*FLASHING A ROM*, *FLASHAGE*, IN FRENCH)

An operation that aim at changing the executable instructions of an operating system and related apps (rom) so that the original os can be changed and updated. Generally done on android phones to modify the version of the os, as well as getting new features and customizations, removing bloatware, altering custom ui.

FLEXIBLE FLAT CABLE (*NAPPE*, IN FRENCH)

A miniaturized form of ribbon cables, which feature many conducting wires running parallel to each other on the same flat plane. Often connects the smartphone motherboard to other micro-components (e.G. Camera).

H

HACKERSPACE

A hackerspace is an open community laboratory where participants can share resources and knowledge. Activities, usually related to digital technologies, often take the form of workshops, presentations and conferences.

Imei (international mobile equipment identity)

A number, usually unique to identify certain mobile phones, as well as some satellite phones. Usually found printed inside the battery compartment of the phone, but can also be displayed on-screen on most phones by entering a code, or alongside other system information in the settings menu on smartphone operating systems.

J

JAILBREAKING (*DÉBRIDAGE*, IN FRENCH)

The process by which the phone's operating systems is modified to remove restrictions and give greater user control over the device.

JUMPERING

Technique through which copper wire is melted onto the board in order to bypass (or 'jump') faulty elements in the circuit.

K

KLUDGE

A workaround or quick-and-dirty solution that is clumsy, inelegant, inefficient, difficult to extend and hard to maintain.

L

LCD SEPARATION

The process by which the separator screen repair machine is used for separating the broken glass from an LCD lens/ touch screen, or for cellphone glue removal.

M

MAGNETIC MAT

Organizer surface for small electronics repair that magnetically secures all the parts (e.G. Screws) while working on a device.

MICRO-SOLDERING

Connecting wires or small-scale components to each other with an alloy made from tin and copper under a microscope.

MOBILE DRYER

An oven-like machine that heat the components to be dried so that they do not undergo any damage such as delamination, internal cracks, stagnation of the tin from the core, etc. The oven also reduces the temperature shock that results from soldering.

MOTHERBOARD (*CARTE MÈRE OU PLATINE*, IN FRENCH)

The main printed circuit board (pcb) found in general purpose computers and smartphone. It holds and allows communication between many of the crucial electronic components of a system, such as the central processing unit (cpu) and memory, and provides connectors for other peripherals.

MULTIMETER

An electronic measuring instrument that combines several measurement functions in one unit. A typical multimeter can measure voltage, current, and resistance.

N

NETTOYAGE/SERVICING

1. Technique for washing and cleaning the motherboard, restoring functions lost through the intrusion of dirt or moisture.
2. Nettoyage logiciel.

O

OBSOLESCENCE

The process of becoming obsolete or outdated and no longer used.

OVERCLOCKING

The action of increasing a component's clock rate, running it at a higher speed than it was designed to run. This is usually applies to the cpu or gpu, but other components can also be overclocked. Overclocking a phone requires root access and adding new code to the internal memory, which generally comes with some serious risks of damaging the device.

R

RE-BALLING

Re-making missing pins within integrated circuits.

REBOOT (*RÉINITIALISATION*, IN FRENCH)

To execute a device's "boot" process, effectively resetting it and causing the operating system to reload, possibly after a system failure.

RECONDITIONING (*RECONDITIONNEMENT*, IN FRENCH)

Replacing or repairing all the parts that are damaged or broken. Refurbished phones corresponds to devices that are returned to the seller (because of some manufacturing and functioning defect) then repaired under full quality check in order to sold them again.

RECYCLING

The process of converting waste materials into new materials and objects. It is an alternative to "conventional" waste disposal that can save material and help lower greenhouse gas emissions. Recycling can prevent the waste of potentially useful materials and reduce the consumption of fresh raw materials, thereby reducing: energy usage, air pollution (from incineration), and water pollution (from landfilling).

RESSOURCERIE

A social recycling facility that collect and sell/share/lend/give various household devices to anyone interested in lowering their consumption of everyday products. Common on continental europe under different names, sometimes couples with a fab lab/hackerspace.

REPAIR CAFÉ

A meeting in which people repair household electrical and mechanical or digital devices, bicycles, clothing, etc. Often organised by and for local residents, sometimes by hackerspaces. Repair cafés are also commonly held at a fixed location where tools are available and where they can fix their broken goods with the help of savvy volunteers.

RETROFITTING

Refers to the addition of new technology or features to older systems.

REVERSE ENGINEERING (*RÉTRO-INGÉNIERIE*, IN FRENCH)

The process by which a device is disassembled to reveal its designs, architecture to extract knowledge from it, or to duplicate or enhance the object.

ROOTING (*ROOTER*, IN FRENCH)

The process of attaining root access—the administrative superuser permissions—to an android device so that the user can remove the limitations that carriers and hardware manufacturers put on the phone and have the ability to perform operations that are not allowed to a normal users, such as deleting system applications and settings, or running specialized applications.

S

SPUDGER. ALSO "SPLUDGER"

A tool that has a wide flat-head screwdriver-like end that extends as a wedge, used to separate pressure-fit plastic components without causing damage during separation. He flat end of the spudger is often used to loosen or release components inside electronics, for example during the replacement of batteries or touch screens for smartphones.

SWAPPING

The process by which a component (battery, sim card, etc.) Is exchanged. "Sim swaps" is also complex form of mobile phone fraud that is often used to steal large amounts of cryptocurrencies and other items of value from victims.

T

TEARDOWN REPORT

A document that provides a summary of disassembling a

technological product. Includes a detailed analysis of technologies and hardware improvements. Also go by many names such as ’work shop reports’ or ’shop findings’. Often useful for learning and performing repair interventions.

THERMAL COMPOUND

A kind of paste that is necessary to create a reliable transfer of heat from the smartphone’s processor to its heatsink to keep its cpu from overheating.

U

ULTRASONIC CLEANING

A process that uses ultrasound (usually from 20–40 khz) to agitate a fluid. The ultrasound can be used with just water, but use of a solvent appropriate for the smartphone to be cleaned of contaminants adhering to substrates like metals, plastics, glass, rubber, and ceramics. This action also penetrates blind holes, cracks, and recesses.

UNLOCKING (*DÉSIMLOCKAGE*” IN FRENCH)

A sim lock, simlock, network lock, or carrier lock is a technical restriction built into certain mobile phones by their manufacturers to restrict the use of these devices to specific countries and/or network. Unlocking a phone can be done by entering a code provided by the network operator, or through software running on the handset or a computer attached to the handset. Usually the unlock process is permanent.

BIBLIOGRAPHY

Allard, Laurence. 2015. "L'engagement du chercheur à l'heure de la fabrication numérique personnelle." *Hermès* 3 (73): 159-67.

Anderson, Chris. 2012. *Makers: The New Industrial Revolution*. New York: Crown Publishing Group.

Auray, Nicolas. 2016. *L'Alerte ou l'enquête. Une sociologie pragmatique du numérique*. Paris: Presses des Mines/ParisTech.

Bagnall, Brian. 2005. *On the Edge: The Spectacular Rise and Fall of Commodore*. Winnipeg: Variant Press.

Bell, Joshua, Joel Kuipers, Jacqueline Hazen, Amanda Kemble, et Briel Kobak. 2018. "The Materiality of Cell Phone Repair: Re-making Commodities in Washington." *Anthropological Quarterly* 91 (2): 603–633.

Berrebi-Hoffmann, Isabelle, Marie-Christine Bureau, et Michel Lallement. 2018. *Makers. Enquête sur les laboratoires du changement social*. Paris: Seuil.

Bessy, Christian, et Francis Chateauraynaud. 2010. "Le savoir-prendre Enquête sur l'estimation des objets." 2010 2 (54–55): 689–711.

Bosqué, Camille. 2015. "Enquête au cœur des FabLabs, hackerspaces, makerspaces. Le dessin comme outil d'observation." *Techniques & Culture* 64 (2): 168–85.

Bosqué, Camille. 2016. "La fabrication numérique personnelle, pratiques et discours d'un design diffus. Enquête au cœur des FabLabs, hackerspaces et makerspaces, de 2012 à 2015." Rennes: Université de Rennes 2. https://tel.archives-ouvertes.fr/tel-01292572.

Braybrooke, Kat, et Adrian Smith. 2018. "Liberatory technologies for whom? Analysing a new generation of makerspaces defined by institutional encounters." *Journal of Peer Production* 12: 1–13.

Callén, Blanca. 2016. "Donner une seconde vie aux déchets électroniques. Économies informelles et innovation sociotechnique par les marges." *Techniques & Culture* 65–66: 206–19.

Causey, Andrew. 2017. *Drawn to See: drawing as an ethnographic method*. Toronto: University of Toronto Press.

Cavalcanti, Gui. 2013. "Is it a Hackerspace, Makerspace, TechShop, or FabLab?" *Makezine* (blog). 2013. https://makezine.com/2013/05/22/the-difference-between-hackerspaces-makerspaces-techshops-and-fablabs/.

Certeau, Michel de. 1980. *L'Invention du quotidien I, Arts de faire*. Paris: Gallimard.

Chipchase, Jan. 2006. "Mobile Phone Repair Cultures." Présenté à Pecha Kucha, Tokyo. http://www.janchipchase.org/fp/wp-content/uploads/presentations/JanChipchase_PechaKucha_RepairCultures_Presentation_vFinal.pdf.

Cohn, Marisa L. 2019. "Keeping Software Present: Software as a Timely Object for Digital STS." In *digitalSTS: A Field Guide for Science & Technology Studies*, édité par Janet Vertesi, David Ribes, Carl Di Salvo, Yanni Loukissas, Laura Forlano, Daniela K. Rosner, Steven J. Jackson, et Hanna Rose Shell, 423–46. Princeton: Princeton University Press.

Crawford, Matthew. 2009. *The Case for Working with Your Hands*. London: Viking.

Dant, Tim. 2009. *The Work of Repair: Gesture, Emotion and Sensual Knowledge*. Lancaster: Lancaster University Press.

Dant, Tim, et David Bowles. 2003. "Dealing with dirt: servicing and repairing cars." *Sociological Research Online* 8 (2): 1–17.

Davies, Sarah. 2017. *Hackerspaces: Making the Maker Movement*. Cambridge: Polity.

Denis, Jérôme, et David Pontille. 2015a. "Beyond breakdown: two horizons of maintenance work." 15-SES-08. i3 Working Papers Series. Paris: CNRS.

Denis, Jérôme, et David Pontille. 2015b. "Material Ordering and the Care of Things." *Science, Technology, & Human Values* 40 (3): 338–67.

Denis, Jérôme, et David Pontille. 2017. "Beyond breakdown: Exploring Regimes of Maintenance." *Continent* 6 (1): 13–17.

Dougherty, Dale. 2012. "The Maker Movement." *innovations* 7 (3): 11–14.

Edgerton, David. 2006. *The Shock of the Old: Technology and Global History since 1900*. Oxford: Oxford University Press.

Fairware. 2016. "Makers, Movers, and Shakers: How the Maker Movement is Crafting a more Sustainable Future." 2016. https://fairware.com/makers-movers-and-a-sustainable-future/.

Finn, Ed. 2017. *What Algorithms Want: Imagination in the Age of Computing*. Cambridge: MIT Press.

Gabrys, Jennifer. 2011. *Digital Rubbish. A Natural History of Electronics*. Ann Arbor: University of Michigan Press.

Galman, Sally A.C. 2009. "The truthful messenger: Visual methods and representation in qualitative research in education." *Qualitative Research* 9 (2): 197–217.

Gershenfeld, Neil. 2007. *Fab: The Coming Revolution on Your Desktop—from Personal Computers to Personal Fabrication*. New York: Basic Books.

Graham, Stephen, et Nigel Thrift. 2007. "Out of order: understanding repair and maintenance." *Theory, Culture and Society* 24 (3): 1-25.

Gregson, Nicky, Alan Metcalfe, et Louise Crewe. 2009. "Practices of object maintenance and repair: How consumers attend to consumer objects within the home." *Journal of Consumer Culture* 9 (2): 248–72.

Harper, Douglas. 1987. *Working knowledge: Skill and community in a small shop*. Chicago: The University of Chicago Press.

Hatch, Mark. 2013. Hatch, 2013 *The maker movement manifesto: Rules for innovation in the new world of crafters, hackers, and tinkerers*. New York: McGraw-Hill Education.

Henke, Christopher R. 1999. "The Mechanics of Workplace Order : Toward a Sociology of Repair." *Berkeley Journal of Sociology* 44: 55-81.

Higginbotham, Adam. 2019. *Midnight in Chernobyl: The Story of the World's Greatest Nuclear Disaster.* New York: Simon & Schuster.

Himanen, Pekka. 2001. *The Hacker Ethic and the Spirit of the Information Age.* New York: Random House.

Houston, Lara. 2016. "Unsettled Repair Tools: The 'death' of the J.A.F. box." In *Proceedings from The Maintainers: A Conference.* Hoboken.

Houston, Lara. 2017. "The Timeliness of Repair." *continent* 6 (1): 51–55.

Houston, Lara. 2018. "Mobile Phone Repair Knowledge in Downtown Kampala: Local and Trans-Local Circulations." In *Repair Work Ethnographies—Revisiting Breakdown, Relocating Materiality*, édité par Ignaz Strebel, Alain Bovet, et Philippe Sormani, 129-60. London: Palgrave.

Jackson, Steve. 2014. "Rethinking Repair. Media Technologies: Essays on Communication." In *Materiality and Society*, édité par Tarleton Gillespie, Pablo J. Boczkowski, et Kirsten A. Foot, 221–39. Cambridge: MIT Press.

Jackson, Steve, Alex Pompe, et Gabriel Krieschok. 2012. "Repair worlds: Maintenance, repair, and ICT for development in rural Namibia." In *Proceedings of the ACM 2012 conference on Computer Supported Cooperative Work*, 107–16. Seattle, Washington, USA: ACM Press.

Jackson, Steve, Ahmed Syed, et Md. Rashidujjaman Rifat. 2014. "Learning, Innovation, and Sustainability among Mobile Phone Repairers in Dhaka, Bangladesh." In *Proceedings of the 2014 conference on Designing interactive systems (DIS 2014)*, 905–14. New York: ACM.

Knorr Cetina, Karin. 1999. *Epistemic Cultures, how the sciences make knowledge.* Cambridge: Harvard University Press.

Korn, Matthias, et Susann Wagenknecht. 2017. "Friction in Arenas of Repair: Hacking, Security Research, and Mobile Phone Infrastructure." In *Proceedings of the 2017 ACM Conference on Computer Supported Cooperative Work and Social Computing*, 2475–88. Portland, Oregon, USA: ACM.

Laet, Marianne de, et Annemarie Mol. 2000. "The Zimbabwe Bush Pump: Mechanics of a Fluid Technology." *Social Studies of Science* 30 (3): 225–63.

Latour, Bruno, et Steve Woolgar. 1979. *Laboratory Life: the Social Construction of Scientific Facts.* Los Angeles & Londres: Sage.

Lave, Jean, et Etienne Wenger. 1991. *Situated Learning: Legitimate Peripheral Participation.* Cambridge: Cambridge University Press.

Lemonnier, Pierre. 1992. *Elements for an anthropology of technology.* Ann Arbor: University of Michigan Press.

Leroi-Gourhan, André. 1964. *Le geste et la parole, tome 1: Technique et langage.* Paris: Albin Michel.

Lévi-Strauss, Claude. 1969. *The Savage Mind.* Chicago University press.

Levy, Steven. 1984. *Hackers: Heroes of the Computer Revolution.* New York: Nerraw Manijaime/Doubleday.

Liefooghe, Christine. 2018. "Les tiers-lieux à l'ère du numérique : diffusion spatiale d'une utopie socio-économique." *Géographie, économie, société* 20 (1): 33–61.

Lockton, Dan. 2013. "Design for repair: empowering consumers to fix the future." *The Guardian*, 2013. https://www.theguardian.com/sustainable-business/design-repair-empowering-consumers-fix-future.

London, Bernard. 1932. *Ending the Depression through Planned Obsolescence.* New York: Auto-publication.

Mahler, Julianne G. 2009. *Organizational learning at NASA: The Challenger and Columbia accidents.* Georgetown: Georgetown University Press.

Marcus, George E. 1995. "Ethnography in/of the World System: The Emergence of Multi-Sited Ethnography." *Annual Review of Anthropology* 24: 95–117.

Mathews, Gordon, Gustavo Lins Ribeiro, et Carlos Alba Vega. 2012. *Globalization from Below The World's Other Economy.* London: Routledge.

Mauss, Marcel. 1950. "Essai sur le don." In *Sociologie et anthropologie*, 179–82. Paris: PUF.

Mauss, Marcel. 2013. "Les techniques du corps." In *Techniques, technologie et civilisation*, édité par Nathan Schlanger. Paris: PUF.

Moricot, Caroline. 2001. "La maintenance des avions: une face cachée du macro-système aéronautique." In *Organiser la fiabilité*, édité par Mathilde Bourrier, 183–99. Paris: L'Harmattan.

Neff, Gina, et David Stark. 2004. "Permanently Beta : Responsive Organization in the Internet Era." In *Society Online : The Internet in Context*, édité par Philip N. Howard et Steve Jones, 173–88. Thousands Oaks: Sage Publications.

Nova, Nicolas. 2020. *Smartphones. Une enquête anthropologique.* Geneva: Metis Presses.

OECD. 2000. "Literacy in the Information Age. Report of the International Adult Literacy Survey." https://www.oecd.org/education/skills-beyond-school/41529765.pdf.

Oroza, Ernesto. 2009. RIKIMBILI. *Une étude sur la désobéissance technologique et quelques formes de réinvention.* Saint Etienne: Publications de l'Université de Saint-Étienne.

Orr, Julian. 1996. *Talking about Machines, An Ethnography of a Modern Job.* Ithaca: Cornell University Press.

Packard, Vance. 1960. *The Waste Makers.* New York: David McKay.

PAYEN Jean Baptiste, Valérie, Nicolas Nova, et Daniel K. Schneider. 2019. "Learning mobile repairing by doing and with others." In *Proceedings of EdMedia + Innovate Learning 2019*, 1320–24. Waynesville, NC: Association for the Advancement of Computing in Education (AACE).

PERNG, Sung-Yueh. 2019. "Anticipating digital futures: ruins, entanglements and the possibilities of shared technology making." *Mobilities* 14 (4): 418–34. https://doi.org/10.1080/17450101.2019.1594867.

PLOKHY, Serhii. 2018. *Chernobyl: The History of a Nuclear Catastrophe*. New York: Basic Books.

PORTES, Alejandro. 1997. "Globalization from Below : The Rise of Transnational Communities." WPTC-98–01. Princeton: Princeton University.

RADJOU, Navi, et Jaideep Prabhu. 2015. *Frugal Innovation: How to do more with less*. London: The Economist.

RATTO, Matt. 2011. "Critical Making: Conceptual and Material Studies in Technology and Social Life." *The Information Society* 27 (4): 252–60.

RENAUD, Clément, Florence Graezer-Bideau, et Marc Laperrouza. 2020. *REALTIME: Making Digital China*. Lausanne: PPUR.

RICHTERICH, Annika, et Karin Wenz. 2017. "Introduction. Making and Hacking." *Digital Culture and Society* 3 (1): 5–23.

RUSKIN, Jérôme. s. d. "Rien n'arrêtera le mouvement Maker." *Les Echos*.

SCHULZ, Yvan. 2017. "Réassemblages marginaux au cœur de la "Mecque du hardware." *Techniques & Culture* 1 (67): 84–99.

SCHWANHÄUSSER, anja. 2016. *Sensing the city: A companion to urban anthropology*. Basel: Birkhäuser.

SENETT, Richard. 2008. *The Craftsman*. New Haven: Yale University Press.

SHRIVASTAVA, Paul. 1992. *Bhopal: Anatomy of a crisis*. Thousand Oaks: Sage Publications Ltd.

SLADE, Giles. 2006. *Made to break*. Cambridge: Harvard University Press.

STAR, Susan Leigh. 1995. "Introduction." In *Ecologies of Knowledge: Work and Politics in Science and Technology*, édité par Susan Leigh Star, 1–38. Albany: SUNY Press.

STREBEL, Ignaz, Alain Bovet, et Philippe Sormani. 2018. *Repair Work Ethnographies—Revisiting Breakdown, Relocating Materiality*. London: Palgrave.

SUCHMAN, Lucy. 2007. "Agencies in Technology Design: Feminist Reconfigurations." https://www.lancaster.ac.uk/fass/resources/sociology-online-papers/papers/suchman-agenciestechnodesign.pdf.

TARRIUS, Alain. 2005. "Circulatory territories." *Mondialisations* (blog). 2005. http://www.mondialisations.org/php/public/art.php?id=18355&lan=EN.

TARRIUS, Alain. 2015. *Etrangers de passage : poor to poor, peer to peer*. La Tour-d'Aigues: Editions de l'Aube.

TSING, Anna. 2015. *The Mushroom at the End of the World. On the possibility of Life in Capitalist Ruins*. Princeton: Princeton University Press.

VAUGHAN, Diane. 1996. *The Challenger launch decision: Risky technology, culture, and deviance at NASA*. Chicago: The University of Chicago Press.

VINCK, Dominique. 2003. *Everyday Engineering. An Ethnography of Design and Innovation*. Cambridge: MIT Press.

VINCK, Dominique. 2007. "Retour sur le laboratoire comme espace de production de connaissances." *Revue d'anthropologie des connaissances* 1 (2): 159-65.

VON HIPPEL, Eric. 1988. *The Sources of Innovation*. Oxford: Oxford University Press.

WEAVER-HIGHTOWER, Marcus, Paul Kuttner, et Nick Sousanis. 2017. "How to draw comics the scholarly way." In *Handbook of Arts-Based Research*, édité par Patricia Leavy. New York: Guilford Press.

WENGER, Etienne. 1998. *Communities of Practice: Learning, Meaning, and Identity*. Cambridge: Cambridge University Press.

Project management: Nicolas Nova
Field research: Anaïs Bloch and Nicolas Nova
Texts: Nicolas Nova
Drawings and field notes: Anaïs Bloch
Photography: Anaïs Bloch and Nicolas Nova
Proofreading and editing: Justin Pickard
Project website: https://head.hesge.ch/mobilerepaircultures
Design: Raphaël Verona and Aude Meyer de Stadelhofen
Printing: MAS Matbaa, Istanbul
Published by IDPURE éditions, Lausanne, November 2020
With the support sof IDP—Independent Design Publications

Acknowledgements: Justin Pickard for his tremendous help on this manuscript. Laurence Allard, Jan Chipchase, Jérôme Denis, Matthieu Duperrex, Benjamin Gaulon, Fabien Girardin, Marie Goyon, Marc Laperrouza, Hervé Munz, Clément Renaud, Maria Roszkowska, Yvan Schulz, Bruce Sterling, Anne-Catherine Sutermeister, Jasmina Tešanović, Mathilde Bourrier and Dominique Cardon for their comments and support, as well as all our interlocutors and all the anonymous maintenance priests/priestness we ran across.

Dr Smartphone is a publication supported by the Swiss National Research fund (SNF Project #170 000, 01.01.2017–31.12.2018) and HEAD–Genève, Geneva University of Art and Design.

IDPURE éditions
ISBN: 978-2-9700992-6-0

FNSNF

Hes·so
Haute Ecole Spécialisée de Suisse occidentale
Fachhochschule Westschweiz
University of Applied Sciences and Arts Western Switzerland

—HEAD
Genève